COUNTDOWN

Aviation safety seems to be measured by fatalities. It takes a disaster to move the authorities, to cut the red tape, to redirect priorities and to establish new procedures.

For instance, here are some U.S. domestic airline statistics to think about.

YEAR	TOTAL ACCIDENTS	TOTAL FATALITIES
1967	70	286
1968	71	349
1969	63	158
1970	55	146
1971	48	203
1972	50	190
1973	43	227
1974	47	467

As this book goes to press another accident becomes front page news. On June 24 an Eastern Airlines 727 jet crashed while approaching JFK airport in New York. It proved to be the worst single plane disaster in U.S. commercial aviation history, claiming 112 lives. This is what a statistic looks like . . . pieces of wreckage, rows of shroud-covered bodies. Will these deaths improve air safety? Is there a better way to fly?

Author's Note

In the year 1974 the U.S. airlines had nine fatal accidents which resulted in 467 passenger fatalities. This is the highest death toll since 1960. Little has been done to improve conditions and unless the airways and airports do something positive the 1974 record will soon be broken.

This book was originally published two years ago in hard cover. However, it has been revised and updated for this edition. Additional material will be found at the end of the chapters in italic type.

Brian Power-Waters
April 1975

SAFETY LAST

by Captain X

NEW AND REVISED EDITION

PINNACLE BOOKS • NEW YORK CITY

To my wife Kay and my daughters
Lisette
Laurette
Linette
Lanette
and to my mother
who helped so much.

Photo Credits

The photographs in this book are published with permissions as indicated here, on a page by page basis, from top to bottom, left to right, according to layout: Opening page; UPI/Compix. Photo insert, page 1; Pan Am, Pan Am. Page 2; all UPI. Page 3; UPI, Wide World, UPI, UPI. Page 4; Wide World, Wide World, UPI, UPI. Page 5; Wide World, UPI, Wide World, Wide World. Page 6; Wide World, UPI, UPI. Page 7; Wide World, Wide World. Page 8; PATCO, PATCO, Tom Hollyman, D. H. Stoops.

SAFETY LAST

A Pinnacle Books edition, published by special arrangement with The Dial Press.

ISBN: 0-523-00715-9

First printing, September 1975

Printed in the United States of America

PINNACLE BOOKS, INC.
275 Madison Avenue
New York, N.Y. 10016

CONTENTS

INTRODUCTION

During the past eighteen months the American public has gained a greater awareness of the Federal Aviation Administration's problems in improving the safety of air travel. The FAA has been taking it on the chin from Congress, the General Accounting Office, and the aviation community. It has had difficulty keeping its promotion of aviation growth separated from its mandate to provide maximum safety. However, there now *seems* to be a more positive FAA responsiveness to the demands for safety. With help from the aviation community (all pulling together) the FAA might—just might—make significant strides in air travel safety. The FAA desperately needs help from everyone. Its more than 50,000 employees and $2 billion per year budget are just not enough to bring order to one of the most complex man-machine systems in the world—the National Airspace System.

A salient point emerges from *Safety Last* which provides the cornerstone for all safety work in aviation. It is aviation's failure to learn from the past. We continue to ignore human factors, good and sound management practices, commonsense applications of safety guidelines, and a too-great reliance upon compliance with the Federal Aviation Regulations—a large set of broadly written, *minimum* standards that do not even suggest good safety performance. So much has happened in the past to guide our future, but we seem to forget—or we fail to communicate.

The Aviation Safety Institute believes, and *Safety Last* shows, that our present unhappy safety record is reflected in our failure to communicate at all levels. We all share in the responsibility for this failure—there are the various aviation associations (special interest or pressure groups), the FAA, the Civil Aeronautics Board, the Administration, Congress, etc.—each trying to further its own interests in aviation safety. Somehow key things don't get transmitted and understood. Thus, we find a diversity of efforts without a common thread of understanding.

A classic case in point was the divergent viewpoints and interpretations revealed during the Flight 514 hearings in February 1975, when pilots, controllers, and the FAA all viewed the problems of landing approaches differently. United Airlines had an immunity program in operation which could have possibly prevented the Dulles crash, but there was no communication between the airline and FAA. You see, the airline had had a near miss with the same mountain a few weeks before the fatal TWA crash. Communications failed and the immunity program failed—we did not learn from past experience.

H. W. Heinrich, in his 1931 book *Industrial Accident Prevention—A Scientific Approach*, stated that ". . . attention is centered upon more spectacular occurrences to the exclusion, in part at least, of adequate consideration of minor accidents. . . ." According to Heinrich, there is no such thing as a minor accident—there are only major and minor injuries. Accidents produce injuries, and one is the result of the other. We see a failure of the aviation community to take advantage of Heinrich's work. Our government has not applied sound principles for accident prevention. (The one possible exception was NASA's moon shot program.) Accidents are controllable; injuries are not. So how do we initiate controls to *prevent* the accident without using the burning wreckage and body count or a smoldering hole in the ground to guess at causes?

We must communicate safety problems as we see them. We

need to report errors that are committed, yet protect all concerned parties. Our interest should not emphasize delivery of recriminations to the offender, rather we should find the errors and conditions that can come together to produce an accident. *This is true accident prevention*. The Aviation Safety Institute has developed an anonymous safety reporting system which uses Heinrich's theories and aviation people's concern for safety problems to pinpoint critical errors and conditions that lead to accidents. The whole process centers around effective communication. People have to report their concerns to an organization which is wholly independent, unbiased, and protective of the reporter. This group must have the patience, persistence, and scientific curiosity to pursue safety problems until solutions are achieved.

The Aviation Safety Institute has, for more than two years, worked in this framework to aid the FAA, airport authorities, air traffic controllers, pilots, and others to correct a wide variety and range of safety problems. We have found that total reliance on the government for our welfare is naive. Involvement of the American private sector is what provides the effective catalyst for reaction to problems. The private sector is sufficiently concerned with air travel problems to demand action.

Regardless where you sit–in the public or private sector–there are some things you can do to help reduce and improve safety. First, read this book. By Chapter 12 you should have formulated opinions of the condition of air travel. Second, tear out and use the safety reporting form at the end of the book. Send ASI your complaints about safety–not about your lost luggage or fouled-up seating arrangements. Third, let your Congressman or Senator know how you feel about air travel problems. Then, in whatever sector of American life you reside, you can be assured that you have contributed to aviation safety, and safety won't be last.

–John B. Galipault
President, Aviation Safety Institute

Chapter One

IS THIS ANY WAY TO RUN AN AIRLINE?

IT WAS a typical August morning in the eastern part of the United States. The humidity was unbearable, and the temperature had already climbed to 83 degrees. The unstable air mass rapidly pushing in from the west was producing beautiful but treacherous cumulus clouds. Aviation weather was forecasting thunderstorms by mid-afternoon, with the possibility of tornados. By 2 P.M. the predicted forecast was a reality. Western New York had already experienced gale winds strong enough to uproot trees.

Not a good day for flying, but as long as the storm cells remained scattered, it would be possible to circumnavigate them in hopes of a fairly smooth flight.

Captain Charlie Nolin had just landed his airliner at the Wakefield, New York, airport. After fueling and boarding the passengers he would be ready for the last leg of Flight 12, to New York City's LaGuardia Airport. The weather en route from Boston was slightly turbulent, and when Flight 12 landed at Wakefield the storm cells were still off to the west and boiling furiously. As Flight 12 cleared the active runway and taxied toward the terminal, Captain Nolin dialed the company on the cockpit radio.

"I'd like a real quick turnaround and a top on the fuel. I'd like to try to beat the storm."

"OK, captain, we'll have you out five minutes ahead of schedule if you like."

"OK, swell," he replied. He cut the engines and reached up and pressed the stewardess button four times, his signal that it is clear to open the cabin doors.

The plane had barely stopped when the ground personnel swarmed about it. Baggage carts were quickly positioned, the fueler was already pumping, and the mechanics had their ladders up checking the engines. It takes a lot of good coordination to service and load an airliner in ten minutes, but with the pressure to keep schedules each man does his job . . . or else.

The airline agent made his usual announcement over the PA: "X Airlines announces the boarding of Flight Number 12 to LaGuardia Airport. Passengers may board through gate Number 4."

The outer fringe of the storm was just starting to shed some light rain on the field as forty-four passengers enplaned in short order. The ground agent and stewardess checked the papers for the proper head count, and the door was closed.

Clear No. 2 signal was given, and as the engine began to rotate the staccato of rain on the roof intensified.

As the agent gave the customary wave-off, Captain Nolin acknowledged and told First Officer Trudo: "Get lowest available to LaGuardia. Maybe we can stay under the cells."

"OK."

The thrust was increased from idle and Flight 12 started to move away from the gate and out to the active runway.

Captain Nolin requested, "Let's have the before takeoff check and tell ground we want to get out of here as soon as possible."

"OK, Charlie." First Officer Trudo began the lengthy check

list, making sure that the numerous switches and knobs would be in their proper position.

He had nearly completed his task when the ground controller called, "Flight 12, you are cleared to LaGuardia Airport as filed. Maintain 3,000 for three minutes, climb to 6,000 feet. Expect higher en route. Maintain runway heading for departure squawk 1,000 low."

The clearance was read back. There must be no misunderstanding as to the route or altitude.

Captain Nolin was certainly a good pilot, and with over 15,000 hours in the air he had seen a lot of weather. He didn't like the looks of what lay in wait for him just a few miles from the runway. He turned to Ed Trudo.

"Tell them we're ready and request an immediate left turn to avoid that thunder bumper."

Wakefield tower replied, "Cleared for takeoff. Left turn approved."

Windshield wipers were placed on high in order to see through the deluge of rain. Maximum power was applied, and Flight 12 was on its way.

Ed called out, "Overhead good, power steady V1." (Up to this speed the takeoff can be aborted.) "V2."

Captain Nolin applied the back pressure on the yoke that would lift them into the air. The aircraft quickly disappeared from view into the murky overcast.

The full fury of one of the worst storms ever to hit Wakefield Airport had arrived: hail as large as golfballs, lightning that would scare even the most experienced pilots, and a shifting wind from all quadrants of the compass. Flight 12 was buffeted violently in its quest for altitude. Captain Nolin and Ed were both handling the controls in a fight for their lives.

Barely 100 feet in the air turbulence smacked the left wing to the ground, and what was once a proud airliner began to crack like an eggshell. It cartwheeled down the runway, spilling out passengers as it lost momentum, finally halting inverted.

Miraculously only twenty-two of the forty-four passengers lost their lives; both pilots were killed instantly. The blame for the crash ran true to form. Pilot error. But *was* it?

On January 17, 1964, Captain Tom Mulroy had a flight he will never forget. The weather was bad, with low ceilings, and visibility down to one mile. Captain Mulroy with fifty-two passengers took off, and after climbing into the weather, he experienced serious power loss in the left engine. He shut it down and almost immediately lost the generator on the other engine. With all alternating power lost he would be without gyros and navigational instruments. His only source of electrical power was his battery.

He made a call to the ground station and waited for an answer. None was forthcoming. Fearing he would deplete his already weakened battery, he turned off his radios. Luckily for all concerned he managed to get below the clouds and began flying with reference to the ground. He skimmed across the treetops and hoped there were no high tension wires to tear him from the sky. Moments like this make a pilot earn a year's pay in a few minutes.

After milling around for what seemed like hours, Captain Mulroy spotted the airport and made a successful landing.

Because of the skill he displayed in handling a situation that could have been disastrous, Captain Mulroy was awarded the coveted Dedalion Award. Each year this award is given to an airline pilot who distinguishes himself by skills that save an airliner from disaster. A formal banquet with speeches and all the trimmings was arranged at San Antonio, Texas, so that Captain Mulroy could receive his well-earned trophy.

Any airline would be proud to have a pilot like Tom Mulroy on its seniority list. Any airline but the one he flew for. In no way did management assist Captain Mulroy in his trip to Texas. They told him they were short of pilots and that he would have to make up the trips he would miss. He went to Texas at his own expense. When he returned he made up his trips as ordered. No mention of his fine performance was released to the press or even written up in the employees' newspaper.

Mr. W. A. Patterson, retired president and chairman of the board of United Air Lines, is one of the finest men aviation has ever known. He treated his pilots as executives, which is what they really are, operating a multimillion-dollar office with five or more subordinates. The following is a story about one of his pilots.

Captain Duescher was in command of Flight 746, a Boeing 720, from San Francisco to Chicago. At 35,000 feet over O'Neill, Nebraska, a light chop was encountered, and Captain Duescher asked for and received permission to climb to a higher altitude. Then it happened. Turbulence pitched the sky giant over on its back and fifty-three people thought the end was near.

Flight 746 was out of control and rapidly reaching for the earth. At 14,000 feet the crew were able to gain control and make a successful landing in Chicago.

President Patterson wrote a personal letter to each passenger of Flight 746. In it he said: "I am sure the experience was frightening, to say the least. If we are to be deserving of your confidence, we must be completely frank in providing you with facts involved in such an incident."

Patterson then related the events that took place at 35,000 feet, praising the crew for a job well done. He added: "I don't know of any experience where a flight officer kept his head, and under such a severe experience, had the knowledge of what not to do under such conditions. We can be grateful to him for the deliberate and cool-headed manner in which he handled and saved such an emergency."

A short time later, UAL President Pat Patterson invited Duescher and his crew to lunch. He once again thanked them for a job well done, and gave each man a sizable check.

All airlines are guilty of some unsafe practices, but some airlines are guilty of them all.

President Barus of X Airlines has one burning desire: to

have flights on time. Now it is understandable to have a scheduled airliner on schedule, but not at any cost. President Barus has many times gone to his airline's dispatch office and given orders that only a disturbed person would give. He has been quoted as saying, "Give the pilots enough gas to get to New York and if they have to hold, let them declare an emergency." Fear is one of this management's greatest weapons.

Fortunately for the traveling public, the "don't rock the boat" pilots are few. In the flying game, there should be none. When a captain discovers a mechanical discrepancy on his plane, he calls maintenance to have it fixed. The first thing maintenance does is check his manual to see if the flight can continue with the mechanical defect. If it can, he tries to talk the captain into taking the plane "as is."

When a captain refuses to do this, the next step in the chain of command is to call the dispatcher. (Federal regulations state that a captain and dispatcher must agree that a flight is safe before it can be released.) Now the dispatcher will try to con the pilot into taking the flight. If he fails, the chief pilot is next in the act. He will try various means to get a pilot to go because, as one chief pilot told me: "The main purpose of pilots is to keep the flight in motion and get the people off the ground."

Unless a pilot stands firm for safety, he will end up in trouble. Unfortunately, a few pilots are "company men." They would fly the crates the airplane came in.

Captain Walt Ames was in the holding pattern in Deer Park, Long Island, waiting for an improvement in the weather at JFK. He called dispatch on the radio, told them that Kennedy was still below limits, and explained that the wind was too strong to try an approach from another direction where the limits would allow a landing.

Dispatch proceeded to issue flying lessons. Which is rather comical. He has had no flying experience, and Captain Ames has 18,000 hours. However, try he must, to get the flight in. Walt put up with the irrational suggestions from dispatch for a few minutes, then told him he was getting low on fuel, and thought they should start thinking about going to Newark, their alternate field. Dispatch began to say something about a PAR (Precision Approach Radar) which they don't even have for the runway in question. Captain Ames promptly turned him off and proceeded to Newark.

On January 17, 1964, Captain McCarthy was in command of Flight 28 from Detroit to New York, with an intermediate stop. On checking with the tower, he was informed that the winds were 90 degrees to the runway, with gusts to 27 knots. The runway was marginal in length and width, and the surrounding land dropped off steeply from both ends of the strip. According to the book he was within limits to land, by a scant 1 knot. Not a safe enough margin, so he decided to pass up the field.

When he told dispatch of his plans, dispatch said, "There are forty people for that stop, and ten waiting on the ground to board." He also said that the wind was within limits. "Why don't you give it a try?"

There is only one way to try a landing, and that is to land. If Captain McCarthy were to land and roll up in a heap, the company and the Federal Aviation Administration (FAA) would be the first to call it Pilot Error.

Needless to say, the landing remained canceled and McCarthy proceeded to an airport where a safe but disgruntled group of passengers deplaned. In my twenty years as an airline captain I can count on the fingers of one hand the times that dispatch had been of help to me. They are far more interested in economy.

Each morning at eight the president of X Airlines holds court at the main executive office. It is called the "briefing." The station managers at every airport in the system are linked together by phone. Now is the time for the chief executive to wreak his vengeance on his subordinates. Especially pilots.

All stations check in and when the last one has acknowledged that he is standing by, the show begins. Many airlines use this system, and use it efficiently to improve operations. X Airlines merely uses it to instill fear in its employees.

All of the previous day's flights from every airport on the system are closely scrutinized, and if there were any delays there had better be a good reason. The following is a brief list of the delays caused by pilots.

Capt. Bryan:	Checked out a faulty engine.
Capt. Burchinal:	Requested a statement in writing that his airplane was airworthy.
Capt. Raduca:	Refused to take a plane with an inoperative fire-warning light.
Capt. Meuris:	Experienced weather that required a reduction in speed for safety.
Capt. Grabovsky:	Failed to make a straight-in approach at an uncontrolled field.
Capt. Coward:	Had a fuel leak checked.
Capt. Pierce:	Wanted his cabin preheated. Temp. in cabin—12 degrees.
Capt. Nolin:	Returned to point of takeoff because of a line squall that extended completely across his course.

President Barus openly belittled these pilots. Very few pilots ever heard the dressing down that these men received. And for what? For doing their jobs as the FAA required.

X Airlines, as all other airlines, used line pilots to perform training and checking. Each time a new pilot was selected to be a regional chief, or instructor, he would be sent to the head office for orientation. He would be told his duties, and that he would be given a free hand. The free hand lasted until management wanted the ego of a certain pilot squashed. What better way to make him cry "Uncle" than to fail him on a proficiency check (the FAA requires two a year)? When the line-check airmen refused to down a man for no good reason, they were replaced.

X Airlines, realizing that the regular line pilots would never do the dirty work expected of them, hired some retired navy pilots with no airline experience whatsoever. President Barus would be killing two birds with one stone: (1) he would have nonunion check pilots who were not on the seniority list (airline pilots adhere to strict seniority system); (2) he would be getting pilots to work for half the pay of a regular line pilot.

Here is a group of men with no airline experience checking pilots with over 20,000 hours of commercial flying time. To get the pressure off to a roaring start, the first captain due for his check just happened to be X Airlines' most senior pilot, Captain Flowers.

Flowers had never failed a check ride in twenty years. He was looked upon by his fellow flyers as the living image of the way a captain should look and act. He had more time taxiing airplanes than his check pilot had total flying time. Nevertheless, his number was up. And when a check pilot is out to fail you, you fail.

Many more captains and first officers would soon fall prey to the wrath of management. It is really amazing the thousands of dollars that were somehow dredged up in order to retrain scores of pilots. Especially from a company that continually cries poverty.

Before the coming of the outside check pilots, about 2 percent of the men failed their checks. After management's favorite team got rolling, that number increased to 33 percent. It is interesting to note that the FAA went along on many of these earlier rides, and that they saw no problem with the work of the Air Line Pilots Association (ALPA) line-check pilots.

The contract between ALPA and the pilots employed by X Airlines allows a certain amount of line flying to be performed by managerial pilots. What follows is a list of some of the mistakes made by the new check pilots when they left the safety on the home-field traffic pattern and tried to play airline pilot:

Mr. Irwin: Failed to check out in a new airplane three times. Landed on the wrong runway over the top of an airliner already in position for takeoff.

Mr. Johnson: On an inaugural flight with a planeload of company officials, landed so far down the runway that he had to be towed backward before he could turn around. Landed at the wrong airport while captaining a chartered flight. In clear weather flew right past his destination airport until questioned by his co-pilot.

Mr. Chahi: Was issued a citation by the FAA for circling an airport below the prescribed circling limits.

This is the caliber of men that management chose to oversee the pilots on X Airlines.

Some of the biggest gripes that management has with its pilots is the fact that they don't all fly like lunatics. They take the time to inspect their planes. They don't always fly at top speed in turbulences, and they don't make schedule at any cost. The airlines spend millions in advertising the fact that their passengers are treated like royalty. They should practice what they preach.

When a company check pilot gives a man a route check (check on a regular scheduled trip), he will grade the pilot for his knowledge of the equipment, and also for the way he flies. Each check pilot also has a classified check sheet that he uses on each captain. This list includes such unsafe items as:

Does the pilot taxi fast?

Does he take off on the closest runway regardless of the wind?

Does he keep the power up even in rough weather?

X Airlines keeps a time graph on all of its pilots. The flyers who continually beat schedule are the fair-haired boys. Captains who do their job recklessly are given official awards. One such captain was given a plaque and a "Well done" from the president. And the captains who fly as professionals are at the bottom of the heap. This type of thing can only lead to competition among many of the pilots, dangerous competition.

How is it that one man can beat schedule, and another can't

make it? Simple! Throw the safety rulebook out the window. Fly without a clearance in marginal weather. Call the tower ahead of another flight, even if you are behind him, in order to get a faster landing sequence. Don't use all the runway for take-off. Land on the runway most nearly aligned with your course. Cancel your IFR (instrument flight rules) even when not clear of the clouds. Taxi as if you were driving in a sports-car rally. And last but not least, mark down the schedule departure time in the logbook and run the engines at high-power settings in order to cover up your lie.

As long as you make schedule, management sits in your corner. If you don't, expect the consequences. Most of the preceding practices are sanctioned by the company. A few are concocted by captains who lack the guts to stand up and fight for a safe operation. Their motto may be "Don't make any waves and retire happy," but they should add, "Providing you live that long."

On June 22, 1967, Captain Snell called the crew scheduler and told him he would be out for a few days with a cold. He would call him as soon as he could fly again. The next day the company phone rang and Captain Snell let it go until they got tired of ringing (most pilots have two phones, one for the company, and one for their friends). About two hours later, a state trooper arrived at the captain's home. Snell was alerted to his presence by his youngest daughter, Laurette.

"Daddy, there's a policeman at the door and he's got a gun."

Snell opened the door and asked, "What's the problem?"

"Are you Captain Snell?"

"Yes."

"Well, your company was worried about you, and they sent me over to see if you were all right."

The answer Snell gave him to relay is unprintable. It just shows how far a company will go in order to fill a pilot's seat and get a trip out.

Take the case of Captain Mergeler, a good solid pilot who flies right by the book, but seldom makes schedule. He went to his FAA doctor and said he was not feeling up to snuff. The doctor recommended he get away from flying for a few weeks and just take it easy. He wrote a note to the airline stating this fact.

The airline was only too willing to oblige. Now they had the opportunity they had been waiting for. Captain Mergeler went fishing for two weeks, and when he returned he attempted to sign in for his next series of flights. Crew schedule told him to contact the chief pilot. He did and the chief told him that they were really worried about his health, and that he would have to submit to a thorough psychiatric exam before he would be returned to the line. ALPA advised him not to take this test. He did, however, go to a FAA flight surgeon and passed his class-1 physical with no waivers.

Captain Mergeler was held out of service for one year. After numerous meetings with his employer and ALPA, no amount of persuasion would make the company back off on their demands. After due process of law a decision was handed down by the arbitrator that Captain Mergeler should be reinstated with full seniority and a year's back pay.

Captain Cisco got started off on the wrong foot years ago, when he failed to make a straight-in approach and circled the field once. This added a few extra minutes to his flight since the airport in question had no control tower, and there were light planes in the pattern, making a circling approach mandatory. Nevertheless, the airline terminated him. ALPA quickly got him back, but the die was cast. He was on the company's list.

On another occasion, while receiving a routine check from one of the check pilots, Captain Cisco encountered turbulent weather. He lowered the landing gear as prescribed by the manufacturers, but the check pilot did not approve of his actions. He was again removed from flight status, and scheduled for a proficiency check, which he allegedly failed. He was given another ride without any benefit of training, which he also supposedly failed. He was terminated by the company on the grounds that he had difficulty passing proficiency checks. At the hearings considerable light was shed on this pilot's past performance.

Captain Cisco had been a captain on the airline for ten years. He had never failed a check ride until the outside pilots arrived on the scene. The only rides he failed were given to him without the watchful eye of an FAA inspector. Whenever the FAA was on board, the company pilot could not pull any shenanigans. During the past ten years Captain Cisco had received four hours and forty minutes of additional training. That would average out to less than thirty minutes' training time per year. The company takes the position that Captain Cisco is a slow learner and requires too much training. Another pilot, one of the Go-Go pets, failed seven check rides and required sixteen hours of training, but he is still flying the line and still beating schedule.

A line captain testifying on behalf of Captain Cisco stated that he was told by the airline president: "The company has no case on the grievant but we are going to take him before a neutral arbiter to teach him a lesson." ALPA won the case and the pilot received all his back pay.

X Airlines is understandably gun-shy of the FAA. With their numerous underhanded operations the FAA is the last person a pilot is supposed to talk to. To prove this point the following is a quote taken from a letter sent by the vice president of Operations to all pilots and dispatchers: "Under no circumstances should you, or do you have the right as an employee, to obtain or act upon an interpolation of a regulation given you by the FAA other than through normal management channels."

It is interesting to note that the FAA issues all pilot's licenses, and has the power to take them away. It all boils down to this: Some companies want their flights to take off at any cost. The FAA says if you take that trip you'll lose your license. It becomes a rather perplexing situation keeping both company and FAA happy without losing your job.

After this brief look at the inner workings of a scheduled pressure cooker, let's look at the background of that crash of X Airlines Flight 12.

Charlie Nolin, the deceased pilot of Flight 12, was not a company favorite. He flew as a professional pilot should, and until August 16, 1963, took no heed of the ever-present pressure. On this day he had the misfortune of having one engine inadvertently go into a reverse pitch just after landing. He tried his best to manipulate the controls but couldn't keep on the runway. He veered to the left, hitting a taxi light which bent two of the blades. There were no injuries in this mishap, but it was at this point that he began to weaken under company harassment.

After the usual hearings he received a letter from the vice president. The letter stated in part that tests were performed on his plane and no malfunction could be found. It continued:

> In summary, you have been the principal in a number of problems which have cost the company a great deal of money and time. These incidents have reflected on your judgment as a scheduled airline pilot and employee. In the future your flights must be handled in a manner consistent with good operating practice and procedures, and in full compliance with company regulations.
>
> In view of your record it will be incumbent on me and the supervisory pilots to carefully monitor the manner in which you conduct yourself and your flights in the future.

A short time after the propeller incident, Captain Nolin was once again on the carpet. He took off from Newark Airport and halfway to his destination was confronted with a massive thunderstorm. Since his radar showed no clear holes in it he decided to return to Newark. (A storm similar to this was the cause of his fatal crash.) He was immediately contacted by the director of flight. He told his superior about the storm and the reason for his return.

The director of flight replied, "If you can't fly the way we want, we'll get pilots that will."

On his next proficiency check ride, the company's chief pilot found Captain Nolin deficient in a number of items and gave him a bust.

During the Civil Aeronautics Board investigation (now known as National Transportation Safety Board) one of X Airlines' check pilots stated that he "was building a file so that they could discharge Captain Nolin."

One hour prior to the fatal takeoff, another pilot from X Airlines called the company dispatcher's office and said, "There is a storm just west of Wakefield covering an area of 20 miles with tops of 50,000. It's a real black one, and it might be worth watching."

That message was received by company but was not reported to any crews. No storm warning was issued by Wakefield tower, but an American Airlines flight that departed five minutes before the tragedy was given vectors to avoid the storm.

ALPA's Region One vice president included the following in his lengthy report to his chief of engineering and safety, Ted Linnert:

> I also met [Mr. Barus] and [V. P. Slicer] of [X Airlines] and at this point I must say that the pilot-company relationship at this accident is the poorest I have ever witnessed. The company is making every effort to close this thing out as pilot error at the earliest possible opportunity.
>
> It would appear that Management will make every effort to insure that the pilot is but a tool to function on demand, and command authority will not be tolerated. A ramp agent at one station had a pilot grounded for two weeks.

The FAA inspector who rode with Captain Nolin on his last check ride said that he was a good pilot. The company check pilot said that he was a "hazard." The official CAB accident report states: "The Board determines that the probable cause of this accident was a loss of control during an attempted takeoff into a severe thunderstorm."

According to ALPA Accident Investigation report, Human Factors Division: "It is a reasonable conclusion that continuous pressure upon the pilot group has the potential of eventually affecting the judgment of a pilot by implying that 'ON TIME' performance is of greater importance than safety."

Was Captain Nolin totally at fault? *You be the judge!*

Chapter Two

THOSE MEN UP FRONT

A BOY'S DREAM, a man's profession, a job that pays up to $81,000 annually for six months' work, two weeks off every month, practically free transportation for the whole family anywhere in the world, thirty days' paid vacation, retirement at age sixty with a pension that exceeds the salary of most men—a job that isn't simply a means of providing for family, but one that can be performed with enthusiasm: That's what being an airline pilot means.

But there are some drawbacks. Very few pilots make it to retirement. If a mid-air, a dangerous approach, or a faulty instrument doesn't get you, perhaps a six-month proficiency check or a physical will.

It's a long, hard road to the coveted left seat of an airliner. Take the case of Jack Banning, twenty-five, from Jason, Montana. Six months before he became a high school teacher, Jack took his first ride in a plane and was bitten by the flying bug. After that, he couldn't wait until his classes were over to jump in his car and drive 50 miles to the nearest airport so he might spend $30 for an hour's instruction. Before he obtained the minimum licenses required by an airline, he spent over $7,000.

Expenses became so great he gave up all social engagements, sold his car and bought a motorcycle. After two years of dedication he completed the requirements for commercial and instrument rating. Now he was not only in the air but walking on it!

Jack Banning submitted applications to most of the carriers; Trans World Airlines was the first to answer favorably. He was told to report to the municipal airport where a pass would be waiting for him to Kansas City, TWA's main training center.

After all the study and sweat it took to get the ratings, you would think the only thing left for him would be a short course from the airline regarding their special procedures, and three landings. Not so. What's acceptable to the FAA is not always satisfactory to the airlines. Each pilot is a potential captain and they want to be certain they can trust his judgment with the lives of hundreds of people, not to mention a multimillion-dollar piece of equipment.

The physical test disqualifies more pilots than any other, so it's the one that's given first. The physical takes over a day and usually reduces the pilot class by 30 percent. Next on the agenda is a battery of questions on general knowledge. The subjects covered include: meteorology, civil air regulations, instrument interpretation, mechanical aptitude, and navigation. Next in line are psychomotor tests which rate your ability to do simultaneous multiple tasks while the examiner has his eye on the stopwatch.

At the training center Jack discovered he was the only pilot who had not been trained by the military, making the competition tougher still. At the end of the morning's exams he was told to report to the chief pilot. Since he was the sole applicant of his group called from class, this made him wonder if the news was good or bad. He soon found out.

"Sorry to give you the bad news," the chief pilot said, and there was real concern in his face. "But you've failed to meet our standards on this morning's exam. I'm afraid we'll have to send you back." He took a paper out of his desk. "Here's your authorization for a ticket back home."

Jack pocketed the ticket and said with feigned cheerfulness, "Is this the end of my two-year sweat?"

The chief pilot smiled. "There are other airlines, you know. All have different requirements."

Jack returned to Montana. The only other offer he received was from a small air-taxi operator. He decided he would rather teach school.

The applicants trained by the military have a decided edge over the ones who learned their flying in a cow pasture. While many civilian pilots, such as Jack, have given up a lot in order to fly, the military pilot does not pay his instructor for lessons.

Airlines can't always get the caliber of trainees they would like. When the demand for pilots exceeds the supply, their requirements drop. Still, many highly trained applicants have been rejected for minor deficiencies. One pilot who already had a captain's rating and many thousands of hours was turned down by Eastern Air Lines simply because he lacked their current height standard by one-quarter inch. He was given this shattering blow by a chief pilot who was so short that when he flew he needed pillows behind him to reach the rudder pedals.

A few pilots have been fired for lack of proficiency as copilots, yet they managed to check out as captains on another line. The right place at the right time has a lot to do with being hired.

For those lucky people who pass all the preliminary exams, their next stop will be ground school. Here every system on the airplane will be studied. A staggering number of temperatures, pressures, charts, capacities, and procedures must be thoroughly understood and memorized. After two months of intensified studies, at last—a chance to fly. Well, at least to manipulate the controls. Each trainee must pass the prescribed course in the simulator.

The simulator is similar to its predecessor, the Link trainer, in only one respect: it never leaves the ground. The size of the simulator corresponds to the front end of a jet. In fact, it is essentially just that. A jet cockpit cut off from the crew entrance door forward. The instructor's panel at the rear of the "box" (as it is called by pilots) has nearly as many buttons and dials as a spaceship.

Here one can simulate any conditions that might be met in flight: engine fires, hydraulic leaks, over temperatures, icing, pressure leaks, and numerous other emergencies can be imitated while safely on the ground. It is so realistic that engine noise can be heard when the throttles are moved. All simulator work is done under mock weather conditions. The cockpit windows are made of frosted glass so that it is impossible to check on your position by looking out. Actual flight training is the last phase of the program. Trainees who have made it through the simulator phase can expect more emergencies in the air.

When graduation time finally arrives and the new co-pilot is assigned to the flight line, his learning has really just begun. He will be flying under the watchful eyes of many captains, and he will be on probation for one year. If he makes it, and more than 30 percent do not, he can count on being given at least one proficiency check each year by a company check pilot. The FAA can spot-check any pilot at any time. Exacting as it is, however, the training of a co-pilot cannot be compared with the demands made on a pilot who is upgrading to captain.

When a pilot is hired he gets a seniority number on the pilot's list. This number literally governs the pilot's life. It completely controls his salary, vacation, hours of duty, days off, domicile, and retention in service should a reduction in pilots become necessary. This democratic system has been in effect on all airlines since the Air Line Pilots Association came into being in 1930. It completely eliminates the necessity of buttering up the boss at a cocktail party; the seniority number you receive when you are hired can be changed only by the resignation or death of any man senior to you.

An airman who starts with an airline as a flight engineer

will stay in that capacity until the airline acquires new routes or purchases new equipment. If the airline expands, he could be promoted to first officer in as little as two years. However, the upgrading from first officer to captain may take anywhere from five to fifteen years.

Once a pilot has a few years of seniority he is more or less married to his airline. If a senior airline captain decided he would like to fly for another company he would immediately be demoted from captain to the lowest number on the co-pilot's seniority list.

Airlines nearly always have more than one domicile for flight crews. Some bases are more desirable than others, and since seniority prevails, only the most senior men can have their preference.

When the time finally arrives for a first officer to check out as a captain, he most likely will have either to move his family to a new base or commute from his present home. There are many pilots who "bid out" (transfer to another domicile) in order to get checked out, and the first chance they get, they return to their old base and home.

You have probably flown on a plane whose crew consisted mainly of captains. Each man who wears four stripes is indeed a captain, but because of seniority only one can act as pilot-in-command. The others perform the duties of co-pilot and engineer and are paid accordingly.

The pilot who wants to have his cake and eat it is the fellow who flies as captain at a base other than his home and commutes to work. Everyone travels a certain distance to work each day, but not quite as far as some international pilots. I know one pilot who lives in New York and is based in California, another who resides in France and works out of Kennedy. A senior pilot can pick his trips so an international crew member may need to report to his base only twice each month. He may be away for as long as six days, but when he returns home he has completed nearly half his required flying time for the month.

A domestic pilot is required to fly between 75 and 85 hours

per month. To many people, such a small number of hours is not really much work. A little-known fact is that in order to get those 75 hours each month a pilot may be on duty as much as 300 hours. Flight planning, layovers, turnarounds, debriefing, etc. all take time.

No other profession requires a man to pass so many tests each year. Once a doctor receives his degree his testing days are over. He may make a mistake during an operation but, right or wrong, he probably won't lose his license.

An airline captain's job is up on the chopping block a minimum of six times a year. He is required to pass two proficiency checks (demonstrating maneuvers he has not performed for six months), one route check, two FAA Class-1 physicals, and one physical given by the company. Between checks, an FAA agent may board his flight at any time to observe his techniques.

Failure on any part of these tests could mean the end of a flying career. In most other professions, the more senior the man, the more he is regarded as a person whose knowledge is indisputable. Not so with pilots. Captains, junior and senior, are all required to pass the same exams.

An airline pilot is continually in training. He has many regulations and procedures to review and memorize. Since no two airplanes fly alike, he must go to school whenever the airline purchases new planes or when upgrading on different equipment.

Scottie Holbrook had been an airline captain for fifteen years. He was flying a prop jet, but would soon be making the big jump from propeller to turbojets. He had just received his notification to report to DC-9 school. Here he would once again be on student status, and many of the habits he formed during 18,000 hours in the air would have to be forgotten. Each night, after the children were in bed, Scottie and his wife, Alma, would go over the flight manual together. She would ask the questions, and Scottie would try to answer them. He got so proficient he practically had the whole manual memorized. Ground school was no problem for him, but flying a jet simulator was something else.

Jets fly so much faster than conventional planes that a pilot must plan far ahead of his aircraft. The acceleration time for a jet engine is much slower than that of a prop, however, another feature that must be learned.

Captain Holbrook finally passed the simulator phase, but to do it he required more time than any other pilot on the line. The actual flying of the aircraft was quite different from the trainer. Try as they may, the manufacturers cannot duplicate the feel of a live plane.

Scottie had his procedures down cold, but mastering the aircraft was something else. At night, when training was over, Scottie tried desperately to relax but each time his tensions grew worse.

"Why don't you go back to the props?" asked Alma.

"Well, for one thing we could use the $7,000 raise, and for another I don't like the idea of a machine beating me. Sooner or later the props will be phased out. Then where am I?"

Scottie finished his flight training and passed his FAA oral exam after five long hours. All that remained now was the FAA rating ride in the DC-9. Both Scottie and his instructor were counting on a good ride. The last flight he was on was good enough to get him a recommendation for his new rating, but his training was hot and cold.

The day of the check finally arrived and from the time he got out of bed until he reached the airport his nerves kept getting worse. He was building up to a great case of "checkitis."

Holbrook was aware that his weakest maneuver was taking off with a dead engine. (One of the engines is brought back to idle thrust just as the aircraft leaves the ground. This simulates a malfunction and at the same time cuts the total power in half.) When he reached the airport he knew that within the hour he would be talking to the company check pilot and the FAA examiner. Sooner than expected, he saw the two men coming toward him, all smiles.

"Hello, Scottie. How about coffee?" asked the check pilot.

"Thanks."

But having sat down with his examiners he couldn't touch a

drop. It was as if he were in a dream. This would be his thirty-first check ride, and with every one he grew more tense. After making the preliminary visual inspection and completing the check list, Scottie told the check pilot to call the tower for taxi clearance. The maneuver he dreaded most, he would have to perform first. Poised at the end of the runway, all that was left was to advance the throttles and hope for the best.

The company check pilot was occupying the co-pilot's seat, and the FAA inspector was riding between them on the jump seat.

"Relax, Scottie," said the check pilot. "Treat this ride as if we weren't here and you won't have any problems."

Easier said than done, with two men watching every move. Before applying takeoff power, a few fears raced through his mind. If he should bust the ride he could only demand one more crack at it. If he flunked again his downgrade would spread like wildfire throughout the company and, worst of all he would have on his conscience the scar of a failure.

"Trainer two, you're cleared for takeoff." The voice from the control tower signaled there was no time left.

Scottie eased the thrust levers forward and she started to roll.

As the air speed came to life the check pilot called out: "Overhead good, engine instruments in the green 80 knots."

At this signal Scottie relinquished the nosewheel steering and put his hand on the control column.

"V1," yelled the check pilot (speed after which a takeoff must be continued) and at the same time rapidly retarded the No. 1 throttle.

The plane immediately lurched to the left. Scottie applied the opposite rudder and fought desperately for directional control. The left edge of the runway was closing in rapidly. He pulled back on the control column. Too late! The plane staggered into the air as the landing gear smashed two runway lights. The check pilot took over immediately, and after a low pass at the control tower for a wheels check, landed without further incident.

After more training Scottie was given the opportunity to try once again. He got by the single-engine takeoff, but did poorly on the air work. He now had two downs. He decided to return to props and stay with them until they are phased out. It might be months or years before they are gone. If the company allows him one more shot at it he must pass or he is out. You never have it made in the flying game until you reach sixty—the age for mandatory retirement.

Pilots are well paid. A prop co-pilot starts at $6,000 per year; the scale for a senior man on a Boeing 747 is $81,000. If the supersonic transports (SSTs) are ever put in service, a pilot's pay will double. Minimum retirement pay starts at about $15,000. Some flyers will get double that amount.

Despite the stress that is ever present in the air, one phrase sums up how a pilot feels about his job: "It sure beats working for a living."

Chapter Three

THE ECONOMY OF SAFETY

ON THE MORNING of November 5, 1969, a Seaboard World Airways DC-8 loaded with cargo and a United Air Lines 737 with fifty-four passengers on board crossed at right angles over the Tannersville VOR ("visual oral range," the most widely used method of radio navigation). Both flights were on instrument flight rules (IFR) and were climbing. Visibility was estimated at one-quarter mile in rain, with neither the ground or sun visible.

United's second officer stated in his report to the National Transportation Safety Board (NTSB): "I shouted to our captain and pointed out the traffic coming up on our windshield."

In reacting to this warning the captain's official report states: "I observed a large jet aircraft on a collision course to ours, converging rapidly. I immediately applied hard-over left aileron and back pressure to the extent of 2 Gs, estimated, the aircraft buffeted, but did not stall."

The Seaboard captain said he faintly saw a picture of the fuselage in front of him through the co-pilot's windshield, but he was so close he couldn't see the wings. As the plane disappeared, he thought it would probably hit the tail. Both Seaboard pilots pushed over hard but felt it was too late to do any good.

If it were not for the quick action of the third crew member, an unfortunate accident would surely have taken place.

In 1948, the Civil Aeronautics Administration (CAA), forerunner to the present FAA, made a ruling that all aircraft 80,000 pounds and over must have a third crew member. This rule remained in effect until pressure was brought against the FAA by the Air Transport Association (ATA) and the manufacturers to renegotiate this regulation. The old weight requirement was a foolish criterion for a third person up front, but at least it was cut and dried. The new method, adopted in January 1964, left the decision concerning extra crew members entirely up to the whims of the FAA. They stated that the complexities of the aircraft, the workload factor, and the degree of automation would be some things to be considered before adding a crew member. On paper this sounded good, but in actuality the FAA gave the green light for a two-man crew on the BAC-111, Douglas DC-9, and the Boeing 737, all new twin-engined jets.

The need for a third crew member on these short-haul jets is greater than on the larger jets, which are required by law to have a crew of three. The increased number of short segments in the flight crews' day accelerates the pace of cockpit duties and increases exposure to other traffic in the lower altitudes and around airports. The concentration required during take-offs and landings far exceeds the hours of boredom that the crews of bigger jets encounter. They sit high in the sky where it is relatively safe; the automatic pilot does the work. The short-haul crews barely have time to relax at cruising altitude, let alone enjoy a lengthy rest while "George" does it. Eighty percent of all jet airline accidents have occurred at low altitudes and near airports; 87 percent of all near misses were sighted by the third officer.

The modern jets that replaced the slow prop jobs do have

better and easier systems; nevertheless, they do fly faster. In the ancient "see and be seen" system that dominates the federal airways, they still pose an operations problem. The BAC-111 has a landing speed higher than either the Caravelle or the Boeing 727, which are its short-haul competitors. The DC-6, DC-7, and VC-10 all have lower stall speeds than the BAC-111, yet they require a three-man crew.

Let's ride the jump seat of a typical two-man jet crew from New York to Washington.

The originating check list calls for forty-six multiple items with many more functions demanded for each one called. The before-starting check requires command and response. The first officer (F/O) calls off the required items and the captain responds to each task. After starting engines there are eight items to check, and while taxiing out fourteen more buttons, levers, and lights must be monitored.

The captain is now taxiing the plane to the runway. While he is listening to instructions from the tower, his co-pilot is running the check list. Now he starts the impossible, but mandatory task of listening to the F/O reading the check list, and making sure that he stays within the confines of the complicated taxi routes instructed by the tower. When takeoff clearance is finally received, the most delicate part of the trip is at hand. During rollout the captain is occupied exclusively with keeping the bird straight and occasionally scanning his instruments.

The co-pilot starts to banter about all the good things that are happening during the takeoff: "Power is good, EPERS OK, overhead panel good, 80 knots, V1, rotate V2."

Now we are in the air and it's time to clean up all the parts that are extended, such as gear and flaps. From now until the final approach the captain becomes almost totally engrossed in his instruments and in monitoring the co-pilot. The F/O calls off the twelve items needed to make the climb complete. While he is performing this function the aircraft is probably at about 1,000 feet and entering "Indian territory" (where all the Piper Comanches, Navahos, Apaches, and Aztecs travel.)

Now is the time to be wary of a collision, but both pilots' eyes are fixed inside the cockpit. Since the co-pilot is only performing one function, it's time to get him on the ball. The tower calls and says to change to departure control. This is done while the co-pilot completes the climb check. With his third hand he sets various VOR radials on his navigation instrument so that the captain can conform to the complicated noise-abatement climbout procedures.

Throughout the entire flight both pilots must maintain a continuous listening watch on two different radios at the same time. One is tuned to the company radio, the other to the man who is trying to keep them from being a fatality, the air traffic controller. As the two radio speakers chatter away and you safely traverse Indian territory, it's time to listen to another radio station.

Each navigational aid is identified by a three-letter Morse code. While identifying the "dahs and dits," the other two fellows are still chattering away over the speakers. The co-pilot is continually changing radio frequencies as the flight traverses different control sectors. On a forty-five-minute flight from New York to Washington he will change radio frequencies sixteen times. This entails turning two separate dials to the appropriate numbers. He will be requested to push the ident feature of the transponder nine times, and sixty-six separate radio transmissions will be necessary.*

The number of switches and instruments that must be monitored are far in excess of the capabilities of a two-man crew. There are over 48 instruments on the forward panel and 30 on the overhead. There are more than 113 toggle switches, 56 on the upper panel, 23 on the forward panel, and 34 on the pedestal between the two pilots. Over 103 levers and knobs can be

* Ident: this button when pushed sends a pulse to the radar controller which positively identifies a target by nearly doubling the size of its radar blip.

Transponder: the airborne transmitter that automatically receives radio pulses directed to it by air traffic controllers. When this device is activated by a controller it triggers the airborne receiver and gives a two-slash blip on the radar scope.

counted up front, not to mention 85 lights and 296 circuit breakers which are placed "conveniently," directly behind both pilots. Totaling all the lights, gauges, and switches that must be monitored in the cockpit, we reach a figure in excess of 675.

While all these functions are taking place, no one is looking out the window.

Approaching Washington a descent clearance is received, and the power is reduced. The seven items on the in-range check are completed and we are back down in Indian territory again. Seven items on the before-instrument check are called off, and about ten miles from the Capitol the five remaining items on the before-landing check are completed.

The captain is still flying by his instruments, and after tuning in the appropriate landing aids, identifying their Morse code, studying the approach plate for such pertinent information as heading altitude and missed-approach procedures, he's ready to do his thing. The F/O has to calculate the total fuel burned off and compute the proper air speed for landing.

The approach controller is issuing conflicting traffic reports throughout the approach:

"Capitol Airlines 12, you have traffic 12 o'clock 3 miles opposite direction." Both pilots search for the intruder, but he is lost in the smog. A few minutes go by. "Capitol 12, I show pop-up traffic at 10 o'clock 3 miles eastbound. No altitude information."

The pilot now has the choice of being steered around the traffic, or simply pressing on and hoping for the best. It is mandatory that scheduled commercial carriers be equipped with weather radar. As the name implies, this radar can only detect the best route around storms. It does not pick up other aircraft. It is a good practice to request turns away from conflicting traffic since there is no assurance that your day won't be ruined by having one of these fellows clobber you. Next time you are on a plane that seems to be banking more than usual, don't criticize the pilot for overcontrolling or being lost. All he is trying to do is to keep you alive.

Now is the time for the F/O to perform some impossible

feats. He is required by federal law to keep a sharp watch for traffic, as well as call out altitudes, airspeeds, and rate of sink. After touchdown, the plane is taxied clear of the runway and told by the tower to change over to ground control. While taxiing in, eleven check-list items must be completed, and when parked, thirteen more.

During this flight a radio contact was made every forty-one seconds. This in itself is enough to keep one pilot busy, discounting all other duties.

You now have a slight idea of what goes on up front. It must be remembered that on this flight all systems functioned normally. Add to the already top-heavy workload a few inoperative items that require a constant watch. To this include an emergency and you can imagine the chance a two-man crew has of coping with it and still keeping clear of all traffic. All that's needed to add a final touch is an FAA inspector in the jump seat running an equipment quiz.

A twenty-four-hour plot of one small traffic sector in Seattle, Washington, showed a total of 746 aircraft below 11,000 feet. Only IFR flights were plotted, and these flights comprise approximately 10 percent of the total traffic a two-man crew must avoid while attending to flying duties.

A simple item such as the automatic cabin pressurization system can be a full-time job if it has to be worked manually. There are planes in the sky right this minute that have as many as five systems or gauges inoperative. Yet the FAA condones this practice.

There are numerous instances when a crew member is required to leave his position and enter the passenger compartment: cabin fires, landing-gear inspection, door warning lights, fluid leaks, strange noises in door areas caused by pressure seepage, and passenger problems that attendants could not handle—illness, fear, belligerence, drunkenness, or violence—may require attention.

When a plane lands at a scheduled en route stop, the two-pilot crew is busy on board preparing for the next flight segment. If there is a third officer he is relieved of this duty and is required to make a visual inspection of the aircraft. The third officer has discovered many dangerous items during this walk-around: fluid leaks, worn tires, burned-out position and beacon lights, and faulty brakes.

A recent situation involving failure of the captain's airspeed, altimeter, and rate of climb instruments just after takeoff was observed by the third crew member. The captain was attempting to correct for the decreasing airspeed by pushing the nose down, when the third man observed that there was a considerable discrepancy between the captain's and co-pilot's flight instruments. By this discovery a potentially unexplainable accident was avoided.

Here is a report from one third officer.

> Suddenly I spotted a light twin-engine aircraft at 12:30 o'clock just slightly below our descending aircraft heading 230 degrees relative to ours. Within a split second I became aware of our collision course and altitude. . . . I immediately yelled to the captain "TRAFFIC TWELVE THIRTY. PULL UP!"
>
> The passengers and crew perhaps owe their lives to the fact that the captain reacted immediately to my warning without question or hesitation. He did not see the other aircraft until after pulling up at which time it passed directly under the nose of our aircraft with less than 50 feet of vertical separation.
>
> This was my probation year and I soon felt more secure that my probation would pass smoothly after the captain turned to me and said gratefully, "Thanks, you have certainly earned your pay for the next five years!"

There are hundreds of documented cases in which the third pair of eyes has saved lives, but the FAA still allows two-man crews.

A big safety item that could shave the workload immensely is cockpit standardization. When an airline buys a new plane it is equipped in accordance with the buyer's wishes. One airline may want its radio control heads mounted on a console between the two pilots, another wants it on the roof panel. Some aircraft have two different types of altimeters, or perhaps dissimilar airspeed indicators. There are vast differences in the same model aircraft ordered by the same airline. One model off the production line will have the No. 13 circuit breaker controlling the autopilot placed behind the F/O. Another production model, supposedly exactly the same, may move this breaker to a position behind the captain, and so on.

There are times when hitting the right button at the right time is vitally important. The switches on a British-built aircraft all operate exactly opposite to those on American planes and there are gauges in some cockpits that are hidden from the view of both pilots, vital warning lights that cannot be seen in a normal sitting position.

Imagine the problems created when one airline buys a plane from another. Since both carriers tailor their cockpits as they see fit, two similar aircraft may have totally different cockpit displays. Many pilots change planes two or three times daily. Each supposedly similar plane may have completely dissimilar cockpits. Not an easy task for the crew members to adapt to the new surroundings. More than one accident has been directly attributed to a pilot's getting out of one plane where the *flap handle* is on the left side of the console, and flying another plane that has the *gear handle* in that position.

For example, on landing rollout a captain attempted to raise the flaps by moving a lever that was in reality the gear handle. The wheels retracted and the plane scraped along on its belly and burst into flames. Ten people died in the ensuing fire. The pilot was blamed, but the guilty party was the FAA.

The FAA has no regulation regarding the standardization of aircraft cockpits. They do state that airlines that interchange flights should have fairly similar cockpits. But no mention is made of standardizing each airline fleet.

Circuit-breaker panels containing hundreds of buttons are found behind the captain and F/O. They are not readily accessible in flight. If one of the breakers should pop out, it would not be seen by either pilot. Overhead panels go unnoticed for many minutes. The crew has enough to occupy themselves with the forward panels and traffic scans.

The third man sitting between the two pilots is in the best position to check these panels at a glance. His job is not concerned with guiding the craft through the air, so he has ample time to check these out-of-the-way breakers.

We can all be thankful that the pilot group of United Air Lines has made a strong stand for a third crew member on their 737s. Other carriers using this machine have followed United's lead and are keeping a close look at the final decision on this issue. The cost of a third crew member on the Boeing 737, an aircraft that can carry over one hundred passengers, has been estimated by airline management to be approximately equivalent to the fares of two passengers.

The Boeing Corporation has this to say concerning the third crew member: "The third man in the cockpit could cost us $100 million in lost sales of the 737."

Mr. W. A. Patterson, now retired as president and board chairman of United Air Lines, acquired a vast experience with the economics of safety. Speaking before the Flight Safety Foundation in Madrid, he said: "No one gave serious consideration to the economics of in-flight motion pictures. They are probably costing this industry $25 million per year. . . . The marketing man tells you this is what the public wants. On none of his questionnaires have I observed whether SAFETY was what the public wanted. . . . If it's a question of whether it's a crew member and you honestly feel that you should have one more, versus motion pictures—put that up to the public by the marketing man and he'll find out what the public wants."

Department of Transportation Secretary John Volpe estimates that the total commercial fleet will increase from 2,452 at the beginning of 1968 to 3,600 in 1980. The general aviation fleet, which numbered 114,186 in 1968, is expected to increase to 170,000 by 1975 and 214,000 by 1980.

The record of near accidents is a warning that if the FAA

continues to certify jet aircraft for airline operators without a standard crew of three crew members, it will increase the risk of tragic accidents substantially.

The third crew member is needed today as much as he ever was since pilots still have to operate on unsafe airways and within a system that expects them to see an on-coming jet and avoid it.

If airlines flew only when most of their important components were working, then and only then would two pilots be a safe operation. But planes today are often dispatched with inoperable autopilots. In fact, there is no regulation requiring that planes be equipped with automatic pilots.

These electronic devices are by no means luxury items. They are essential, especially on two-man jets, since they allow a pilot to take a breather and look outside to see what traffic he may be running into. Without the autopilot, one member of the two-man crew is lost for traffic scanning purposes because he must devote his time to manipulating the controls.

Captain John J. O'Donnell, president of the Air Line Pilots Association (ALPA), staunchly supports having three members on every airliner.

When the short-haul jet was introduced some ten years ago there were far fewer aircraft in the sky. Today, with some 1,400 aircraft being sold each month–not to mention more airliners and military craft–the traffic is even more congested. In fact, the airways have long since passed the saturation point.

The record plainly shows that two-man aircraft experience more near misses as well as actual hits than do planes with an observer on board. Not only can the third man spot traffic more efficiently because of his overall view, but he is a comforting aid in all of the following circumstances: unruly passengers, fire in the cabin, passenger evacuation after a crash. He has also proven to be of invaluable assistance in hijacking emergencies.

As long as we have to operate under a dangerous mixture of VFR and IFR and with aircraft components that are inoperative, a third crew member is not a whim of ALPA—it is instead one of the plane's best safety devices.

Over 90 percent of all the navigation aids have a Morse code identifier. This means all pilots must be up on their "dits" and "dahs." More than once pilots have tuned in the omni station in order to get to the destination airport and then have forgotten to change to the ILS. The two code identifiers are exactly the same except for two dots that always precede the ILS identifier. Very few omnis have a simple, foolproof means of identification and that's by voice. When these stations are tuned in, a continuous tape gives the name of the station. Which seems easier? "Dit dah dit, dah dah dah, dah dit dah dit" or simply a voice saying Rochester omni?

There are numerous airports without approach lights as well. These extremely high-powered lights extend outward from the end of the runway for 3,000 feet, and they are supposed to have five positions from dim to full bright. However, there are a few towers where the lights have only three positions. Some airports don't even have the full complement of lights, but there are no guidelines on the approach charts to warn the pilot that the lights on a particular runway may only extend 1,500 feet.

The few runways that have an overrun and underrun usually void their existence. This valuable piece of pavement at each end of a runway can come in mighty handy if you should happen to touch down short. However, nearly all of these strips have runway threshold lights that traverse the runway ends. Pity the pilot who misjudges his touchdown point. Granted it is better than landing short in the dirt, but you will feel just as bad when you scrape your wheel rims on the concrete after having blown all the tires out while rolling through the runway threshold lights.

Many facilities have switching problems with the runway lights. If the lights are up too high and you request a lower setting, the lights may go off completely before the new selection takes hold.

Approach lights are of such importance that when they are inoperable the landing limits increase a quarter of a mile. It is therefore absolutely necessary for the pilot to know if they are functioning or not before he begins his approach.

On July 31, 1973, a Delta Airlines DC-9 was on final approach at Boston's Logan Airport. There were radar vectoring discrepancies, and the cockpit configuration was new to Captain Streil who had recently moved from Northeast to Delta. The weather was varying from limits to below, so the tower operator was unable to see the instrument runway. Within the next few minutes eighty-eight lives were lost because of the system.

The crash wiped out the approach lights and triggered a warning in the tower. But because of the numerous false alarms registered by these lights, the controller simply switched off the alarm and went about his business. Two other flights were cleared to land, but luckily *they failed to break out of the clouds in time and executed missed approaches.*

An alert construction worker saw the crash and drove to the base fire station to report it. The tower controller was first alerted by the driver of the fire truck who called the tower for permission to cross a runway. When asked where he was going, he replied, "There's a crash out there."

The faulty alarm was not wrong this time. But many times prior to this disaster there was a sign placed on the approach warning light stating that if the light came on it should be ignored. This is contrary to all Federal air regulations.

Five months later the powers that be at Logan Airport contributed to another crash, but this time no one was killed.

On December 17, 1973, Iberia Air Lines flight 933, a DC-10, was approaching the northwest runway 33 L. Once again the weather was close to limits and the runway breaking action was reported as fair to poor. When Cap-

tain Gaztelu broke out of the clouds he realized that he was perilously close to the approach lights. He immediately eased back on the yoke and applied power to break the sink, but it was too late. The DC-10 sheared off the right gear and skidded to a stop some 3,000 feet to the right of the runway. This time the approach light monitor was heeded and the crash trucks were promptly dispatched.

There were two factors contributing to this crash, either of which by itself would have been sufficient. The first was a windshift on final approach that as yet cannot be forecast. This meant that the DC-10 would be traveling faster over the ground than usual because of a tail wind. To correct this the pilot must increase the descent angle. This natural phenomenon coupled with a faulty glide slope transmitter could only spell disaster.

The majority of today's airports were not designed for jumbo jets and Boston is no exception. The location of the glide slope transmitter and the width of many taxiways are fine for all but the largest jets.

The glide slope receiver in the DC-10 is located in the nose section, but it should have been placed nearer to the wheels to allow for the plane's size. The glide slope transmitter at Boston was imporperly located (and to date it is still in the same location). It should have been moved further down the runway to give big jets margin for error.

On February 24, 1972 (a year before the above crash), the FAA issued order 8260.24. This directive established the relocation of glide slopes to allow for the safer touchdown clearance of wide-bodied jets. The Boston glide slope facility was not located as the FAA had decreed. Lack of funds was the excuse. Because of the greater distance between the top and bottom of the DC-10 as compared to the DC-8 the margin for error is greatly reduced. A DC-8 has 30 feet between the glide path beam and the runway threshold as compared to the giant DC-10 which has a scant 7.8-foot clearance. This minimal distance can only be enjoyed if the pilot is dead center on course. It becomes in-

creasingly difficult to be in the groove with tail winds, gusts, etc. Yet the FAA expects a pilot to do this job consistently with only 7.8 feet between him and disaster.

The reason I chose Boston from among the many other airports is because of the two incidents that occurred there clearly illustrate what goes on at a major air terminal. Many other strips throughout the world would have served as well since they are not even as good as Logan.

Chapter Four

"IT'S UP TO YOU, CAPTAIN"

An airliner's logbook is interpreted many different ways by many different people. In this book the exact times of each flight are recorded; the operating times of the airframe, engines, and components can be checked; and mechanical discrepancies reported by pilots and mechanics are all there in a permanent record.

A few captains breeze through discrepancies in the aircraft logs. Their philosophy says that if it flew in, it can fly out. True, it would be economically unfeasible to have mechanics and parts at every stop an airline makes. Only the larger cities have maintenance bases, and service checks and overhauls are conducted at the main base.

But what happens when a faulty plane lands at a base where there is no maintenance? If it were not for the minimum equipment list, the MEL, the plane would be grounded and mechanics would be flown in to repair it. The MEL is written by the aircraft manufacturers with the approval of the FAA and the blessing of the Air Traffic Association (ATA), an organization to which all scheduled airlines belong.

An aircraft manufacturer tries to design a safe, efficient, economical airplane with as many fail-safe features as possible. The ATA seems mainly interested in how many safety items the airplane can fly without. The MEL contains a listing of the various components that can be ignored in order to "keep the flight in motion." Here is an example of the MEL in action.

On January 18, 1969, a United Air Lines 727 cargo liner, Flight 266, departed Los Angeles International Airport for Chicago. Shortly after takeoff the plane crashed in the ocean, killing the crew of three. The captain was operating under the MEL which stated that one generator could be inoperative as long as the other two were functioning. After liftoff one of the engines had to be shut down. This left only one generator to supply the whole electrical load. It failed. A fatal crash resulted.

Consider the Air Line Pilots Association information bulletin of March 19, 1969, concerning flying with a generator inoperative: "The operation of an aircraft with an inoperative generator was labeled 'unsafe.' "

It was pointed out in the preliminary findings of the National Transportation Safety Board on the January 18 accident involving United Air Lines Flight 266 that the accident would not have happened had the FAA checked ALPA's warning about "inoperative generators."

The letter also noted: "Thirty days have now elapsed since the accident and no action has been taken by the FAA or the ATA to solve the problem.

"Consequently, ALPA recommends all pilots not to accept aircraft with an inoperative generator, as it is considered by the Association to be unsafe."

The latest copy of United's MEL, dated January 15, 1971, two years after the accident, still allows a 727 to depart the ramp in scheduled service with a generator inoperative.

The captain who refuses a flight because of the aforementioned safety item would be open for disciplinary action from the company.

All maintenance bases have foremen, and most foremen do

a good, honest job. Nevertheless, a few of these men will sign off an airplane as airworthy when it is not. These men have only one object in mind: get the plane on the gate for a scheduled departure, ready or not.

Take the case of Captain Steve Argyris. He boarded his flight the customary twenty minutes before departure and began to check the squawks in the logbook. Turning to the remarks on the preceding flight he spotted an item that interested him. He called a mechanic and they discussed the merits of a write-off. (Each mechanic signs for the work he performs on the aircraft.)

"Main hydraulic line near Y fitting failed. Repaired with a temporary flex line. Should be replaced as soon as possible," the report read. The mechanic stated that he did not work on the plane personally. It was signed off by the maintenance foreman. He said it was not leaking now and it seemed OK to go.

Eight miles out on final approach at New York's Kennedy Airport, this hydraulic line ruptured, leaving the crew with no flaps, no windshield wipers, no brakes, and no nosewheel steering. The landing gear would fall under its own weight. The landing was uneventful, and after coming to a stop the captain called for a tug to tow them to the terminal.

The next day Captain Argyris and F/O Grimes were ordered to their airline's headquarters. Here they were thoroughly questioned as to what procedures were used and whether their duties were performed properly. There was no mention of a job well done, neither was any action taken about the temporary repairs that had caused the accident.

Captain Argyris tracked down the men who had worked on the plane. They told him they did not have the proper rigid tubing for a good fix so they had used a nonstandard piece of flexible line. When the job was completed they had both agreed that the bend in it was too great and that with 3,000 pounds per square inch of pressure it would surely blow again.

They had refused to sign their name to this work as they did not feel it was airworthy. (If this open item were not signed off, the ship would not depart on schedule.) The foreman, more interested in his job than in safety and with no union to stand behind him, had done his duty and signed it off. He and the crew had been lucky this time. How long would luck hold out?

The FAA cooperates with pilots in regard to reporting near misses. Why, then, is there no reporting program set up for aircraft mechanics? Under the present situation, if a mechanic were to report to the FAA that he was ordered by his company to sign off an unairworthy item, his airline would probably fire him. This is one of the major stumbling blocks in aviation maintenance today. The FAA offers no immunity to the man who reports a potentially dangerous situation.

Here's another case. First Officer Brown had just made a walkaround, outside check of an airplane. He informed his captain of a fuel leak under the left wing. The plane had just landed, a through flight with forty-six people still on board. Captain Murphy and a mechanic checked the leak. It was dripping around the edge of a tank inspection plate.

Maintenance told the captain that they would have the leak repaired in no time.

Captain Murphy exclaimed, "Just how are you going to fix it? With chewing gum?"

"Not quite that bad," the mechanic replied. "We have a special gunk stick that should seal it up."

That didn't sound like too safe a fix, so Captain Murphy decided to check the maintenance manual himself. The book backed up the mechanic's statement. But the captain still wasn't satisfied.

"Say, Bob," he asked the mechanic, "just suppose that stuff seals the leak on the ground, what's going to happen when the wing flexes in flight?"

"All I can tell you is the book says it's OK. It's up to you, captain."

Captain Murphy walked over to Operations and called the chief pilot. He thought it would be better to skip the usual chain of command and start at the top. He told the chief he didn't like the fix for a fuel leak and that he wouldn't take the plane with such temporary repair.

The chief mentioned something about the book, but made no impression upon Murphy. The captain told him he wanted to try the leak in flight. He would like to deplane the passengers and take it up for a test hop. If the leak remained dry he would continue the flight.

The chief agreed to his terms and the test hop was flown. When they shut down the engines and checked the leak it was worse than before. The flight was canceled and, this time, no action was taken against the pilot.

On January 20, 1969, Captain Fred Gibbs and First Officer Jack Ward were the crew on Flight 11, a jet trip from New York to Boston. As the power was applied for takeoff, there was a considerable lag in the spooling up of the right engine. Near maximum power was obtained, but as the flight progressed the No. 2 engine continued to act up.

After landing in Boston the captain wrote up the malfunction and also told the mechanics the difficulties he had encountered. The plane was taken to the maintenance hangar and after about two hours returned to the gate to continue in service. Captain Gibbs queried the maintenance department concerning the work they performed. He determined that the problem was not completely eliminated and refused to fly the plane. It was returned to the hangar a second time.

Once again the ship was taxied to the gate. After interrogating the mechanics concerning their repair work, Captain Gibbs still felt the plane was not airworthy. Pressured on the phone by the chief pilot and the director of flights, Gibbs remained firm in his convictions and would not fly the plane.

The chief pilot decided to set up a test hop and he called First Officer Ward on the phone. "Mr. Ward, you and I will be going on a test flight. Take care of the flight plan and walk around and I will be over to Operations in a few minutes."

Ward replied: "It seems that as long as Captain Gibbs refuses to fly the plane, it is not proper that I should do so."

Dumbfounded, the chief pilot replied: "Mr. Ward, you should not have answered the way you did because this puts me in a box. I'm about to go over to Operations and order you to go with me on the test flight. If you refuse, you're fired. I'm not trying to threaten you, but you had better learn who's boss. Just say 'Yes, sir' and proceed as instructed!"

In spite of the tongue-lashing from management, neither pilot flew the plane and no immediate action was taken against them.

Nevertheless, the slow, steady pressuring that First Officer Ward was to experience can by no means be taken lightly. Jack Ward had been with the company nearly three years. The greatest part of that time he was on stand-by. (A pilot must be near a phone for twenty-four hours a day as long as six consecutive days in case he may be needed.) Until this disagreement Jack had felt little pressure from the company. Now he was called for the worst trips on the line and usually on his days off. At some airlines days off are meaningless. If they call you, you fly or run the risk of being terminated.

In order for an airline to qualify as a certified maintenance station they must employ at least one licensed mechanic on each shift. This means that many of the men who work on the planes have no FAA rating. As long as the work performed by the man is supervised and signed by an A and P (aircraft and power plant) mechanic it is legal even though it is nearly impossible for one man to oversee and inspect the work of the many men under him. But so long as he signs it off, the FAA says it is all right.

On June 28, 1965, Captain Charles H. Kimes was in command of Pan American's Flight 843, a Boeing 707 from San Francisco to Hawaii. The plane was serviced with over 14,000 gallons of jet fuel, and the logbook was clear of all discrepancies. The manifest showed a total of 143 passengers, and Kimes would be assisted by a crew of nine. At 2:09 P.M. Pacific Daylight Time takeoff clearance was received. Kimes pushed the throttle and the 350,000-pound giant started to roll. As rotate speed was called out he applied back pressure on the yoke. One hundred forty feet behind the cockpit the giant elevator moved slowly upward and Flight 843 began its climb. The large pointer on the altimeter was just approaching 800 feet above the ground.

BANG! No. 4 engine exploded and fell to the ground, taking with it 25 feet of the wing. The right outboard fire-warning light began to flicker and another explosion was heard. This time the right reserve fuel tank ruptured, ripping more of the underservice of the wing.

Fuel shut-off valve for No. 4 engine was closed and Captain Kimes took a few moments to evaluate the situation and check the flight characteristics. He was able to maintain control and elected to land at Travis Air Force Base. This decision was based on a partial hydraulic failure, the greater length of the runways, and the superior fire-fighting apparatus. (More on this in a later chapter.)

The landing gear was lowered by emergency means and the flight landed at 2:34 P.M. with no further difficulty. Because of the skill of Captain Kimes and his crew, a twenty-five minute nightmare had a happy ending.

On June 25, 1965, three days before the incident, the No. 4 power plant of Flight 843 was overhauled by Pan Am in New York City. It was secured to the No. 4 pod of the 707 and worked perfectly for thirty-nine hours.

It was determined that during periods of takeoff power the third-stage turbine disk and the inner sealing ring expanded and rubbed against each other. Since this only happened at high-power setting, it took numerous takeoffs before it reached dangerous proportions.

A review of the carrier's engine build-up procedures was conducted and depositions taken from the men who assembled the engine. The low-pressure turbine shaft was installed twice to determine the correct size-positioning spacer. The spacer actually determines the position of the low turbine assembly.

Measurements of this spacer were not recorded as they should have been. The mechanic on the shift went home and the next man who took over had no written references to follow. He removed the rotor and then installed it, using the wrong type of tool.

The CAB accident report stated:

> The inspector on duty during the reinstallation of the turbine rotor testified that *he signed off work he had not inspected.* He also lined out a mechanic's initials placed on the work card to indicate the work had been accomplished on the previous shift. He later voided this line out when the work was re-accomplished by a different mechanic.

This procedure is highly illegal.

The official CAB report states:

> The Board determines that the probable cause of this accident was the failure of the third-stage turbine disk. The failure was caused by a transient loss of operating clearance between the third-stage turbine disk and the third-stage inner sealing ring. This loss of clearance resulted from a combination of improper turbine motor positioning during engine assembly, the use of servicable worn parts, and an operating clearance which was less than predicated in design analysis.

Luckily for both you and me not all airlines put schedule first and safety second. The following is the potentially unsafe gamut that takes place when a plane is ready for a maintenance check.

The company inspector writes out a work sheet which includes all the pilot and mechanic squawks.

The foreman checks this sheet and sees how many items he can skip in order to keep the man hours down to a minimum.

Each mechanic is assigned specific duties.

The inspector checks the work after the mechanics have finished and decides if the plane is airworthy. If he is not satisfied with any of the maintenance work, he may not sign it off.

The chief inspector, who has the last word, can and does sign off work in order to have the aircraft ready for a scheduled departure.

The FAA is authorized to monitor any of the maintenance procedures. FAA inspectors have the power to ground an airplane, but they seldom do. In the United States there are approximately 2,500 airliners and 130,000 general aviation aircraft. With only 300 FAA maintenance inspectors, the FAA cannot properly service the maintenance practices vital to safety.

FAA maintenance inspectors work five days a week, eight hours a day. This allows two shifts when there are no government men to check repairs, not to mention weekends, when anything goes. Each FAA maintenance inspector is required to check only four aircraft logbooks per month, and then only the last two weeks' discrepancies. This means that many highly important safety standards can go undetected for months.

On April 27, 1970, Mohawk Airlines received a letter from the FAA stating that they had violated certain federal aviation safety regulations. The following are a few discrepancies.

1. The airline was supposed to have modified the fire extinguishers on five of their planes before a given time period. This was in fact exceeded by over 334 hours on each aircraft.

2. Airworthiness directive 67-25-2 required inspection of the flap drive screw jacks on BAC-111 jets at intervals not to exceed 600 hours. On two occasions jets were flown in excess of that time by 200 hours.

3. Airworthiness directive 67-12-4 requires inspection for

cracks in the rudder horn assembly on the Fairchild 227. Over eight aircraft were flown in excess of this mandatory time from a low of 256 hours to a high of over 1,000 hours.

4. Mandatory periodic checks of navigation equipment are necessary to ensure proper tolerances. The test equipment used to make this check must also be aligned. A check of this testing equipment was long overdue. Various navigation equipment was checked out on test boxes that were past the mandatory date by seven months. One of the most vital pieces of equipment on the plane, the instrument landing system (ILS) exceeded the mandatory check by two-and-one-half years.

The final page of the FAA letter had this to say:

> Under section 90I(A) of the Federal Aviation Act of 1958 as amended, the company is subject to a civil penalty not to exceed $1,000 for each day of each continuing violation of a Federal Regulation. After full consideration of all circumstances concerning both violations we would be willing to accept an offer in compromise of $50,000 in full settlement of this matter.

After haggling over the fine, Mohawk got it reduced to $20,000, which they gladly paid. Why did the FAA refuse $50,000 in lieu of a $20,000 "compromise?"

Here is what hard-hitting James Sparling, Director of Safety and Standards of the Aircraft Mechanics Fraternal Association, wrote in his petition to FAA boss Shaffer:

> It is the Association's opinion that the carriers will continue to flaunt the regulations, with commensurate sacrifice to public safety, knowing full well that the FAA will not aggressively fine the carriers for proven violations. It would appear that it is economically feasible for the carriers to operate unairworthy aircraft in revenue service until they are caught by the FAA as opposed to properly placing the aircraft out of service until required maintenance inspection is completed. And even then, when the car-

riers are caught they know that a "compromise" can be worked out.

When a mechanic at one of the smaller stops is confronted with a complex maintenance problem, he calls the maintenance chief at the main base for technical assistance.

"Hello, Bob, this is Jim Silver at Cleveland."

"What's the problem?"

"Well, we have 44A here with the No. 1 pneumatics system inoperative. I know it's not a MEL problem, but the captain wants it fixed."

"Who's the captain?"

"Howard Steel."

"Oh, that guy'll never take it. Do what you can to try and get him to buy it. If he won't, call me back and we'll set up a delay."

Many logbook squawks are written off with such ever-popular phrases as:

"Couldn't duplicate condition on the ground. Request further in-flight checks."

"Parts NIS [not in stock]."

"OK to continue to service check."

"OK to continue until RON [remain overnight]."

And last but by no means least:

"OK to continue per MEL, ATA chapter 23, page 03–04.06.

The airline takes the trouble to route a plane that needs work so that it will be flown only by a company man. In this way maintenance will almost surely be overlooked and they will get an extra day's service from it. Most mechanics would be glad to repair the maintenance squawks of the pilots. The company will not allow sufficient time for the work to be done. Many of the squawks simply get lost in the back pages of the logbook.

If the FAA were more critical in their spot-checking of these books, there would be fewer aircraft accidents. There is no place in aviation to pit personalities and schedules against safety.

Here's another example of maintenance malpractice. Captain Harvey Trout was the master executive chairman (the ALPA leader for his airline) of his company for a number of years. He was elected to this position by his fellow pilots. The higher a pilot climbs in the union, the lower he becomes in the company's eyes.

On May 2, 1967, Captain Trout reported to O'Hare Field in Chicago to fly his regular scheduled trip to New York. Most captains assign the preflight walkaround inspection to their first officers, but Harvey always made the first inspection of the day himself. When he checked the rear emergency door of his DC-9 he noticed that it lacked the mandatory emergency-exit markings. The plane was a new addition to the fleet and the painters had evidently forgotten the required markings.

Captain Trout refused to fly the airplane. Flying without the proper FAA markings would be in direct violation of federal law. Captain Trout was removed from the flight schedule, and on May 3, 1967, was ordered to the airlines head office.

"Although it was established that this was not to be a 'hearing' and that the sole purpose was to gather facts," he relates, "the meeting fast changed into a harassment session."

Nevertheless, Harvey Trout was returned to the line and was paid for the trips he had missed. Captain Trout's company, X Airlines, was experiencing maintenance problems with a very important component of the controls system, the yaw damper. This device is used as the name implies, to dampen out the side-to-side motion that is ever present in swept-wing jets. Without this device, jets have been known to experience severe control problems in turbulent weather.

X Airlines is the only scheduled airline in the United States that is deactivating all of the series yaw dampers. This is being done with the full approval of the FAA. It seems it is just too much of a maintenance problem to keep the dampers operating. On Saturday, August 2, Captain Trout was assigned a plane with an inoperative series yaw damper. The dispatch release stated that "the autopilot could be used in damper mode only" (which would allow modified dampening).

During the first leg of his trip Captain Trout experienced moderate turbulence. He turned on the damper switch but got no reaction. As soon as he made his first landing he contacted maintenance and told them he wanted the damper repaired as the remainder of his flights would take him into turbulence.

Over an hour elapsed and no maintenance was performed on his plane. Captain Trout did, however, receive numerous phone calls from the airline's head office, all trying to get him to go with the faulty plane. Dispatch finally canceled the flight and Captain Trout deadheaded home. That evening he received a call from the chief pilot.

"Captain Trout, due to your lack of consideration for the inconvenience caused numerous passengers and your total disregard for the *economic impact* upon the company for this action, you are removed from line flying. This removal with loss of all pay for a period of fifteen days."

FAR (federal air regulations) 121.597 and 121.663 make it crystal clear: "Pilots in command have not only the right, but indeed the obligation to decline any flight operation which in their best judgment cannot be safely initiated or completed."

Next time you are on a flight that is delayed for mechanical reasons, ponder the following points:

1. The captain wants to get there as much as you do.
2. The trip you are on turns around at its destination and the pilot flies it back. If he doesn't get there, other equipment is substituted and the captain loses flight pay.
3. If the delay is excessive, the trip may cancel. More loss of pay.
4. If a flight is canceled, the pilot may not return to his home and family as scheduled.

5. The captain is certainly not making any points with his company.

For the passenger, a delay is usually an inconvenience at worst. You should be glad that your captain has one paramount objective: *Your Safety*.

Chapter Five

CAPTAIN LUCKY

ED QUINN was born and raised in a lower-middle-class section of Brooklyn, New York. He fell prey to the flying bug at age nine and spent the remainder of his life preoccupied with aviation and everything it entailed. When his bedroom bookcases no longer had space for his large fleet of airplane models, he strung them on wires from the ceiling.

Slightly built and nonathletic, Ed made aviation his sole extracurricular interest. After high school he worked in a grocery store, and if schoolwork allowed, he set pins in a bowling alley. But weekends were spent at Brooklyn Air Park, a local flying field. Whenever Ed had saved up eight dollars (the cost of a half hour's flight instruction) he hopped on his bike and raced out to the airport.

At sixteen he soloed. Steadily increasing his time in the air, he reached the coveted forty-hour mark. Having already invested over $600 for lessons, he was ready to try for his private license.

World War II interrupted Ed's civilian flying, but it gave

him a chance to be an army pilot. In June 1941 he took and passed the battery of tests that would enable him to wear the uniform of an Army Air Corps Cadet. While the idea of being shot at didn't appeal to him, at least he would be paid for flying. The going was rough in ground school. He barely passed the final preflight exam. He was sent to a field where he would be taught to handle a plane similar to the one he had flown as a civilian. His progress at flight school was above average until he reached the second plateau, the BT-13. Now he would be manipulating a plane with four times the horsepower he was used to, and an instrument panel that would require more skill to interpret.

The first few hours in a new plane are always demanding. No two planes fly alike and it takes some time to get the feel of a new one. When Ed had the maximum prescribed time before solo his instructor warned him that his progress was not satisfactory and that he was going to put him up for a check ride with the flight commander.

Ed was now in the primary stages of checkitis, a nervous condition that grabs some pilots before a check ride. By the time he was scheduled to take his final flight he barely knew his own name. He washed out of pilot training and was reassigned as an air gunner.

Ed was discharged in 1945 and he was naturally disturbed at what the military had done to his flying career. He still believed that flying was his career. Returning to Brooklyn Air Park he found a job working in Operations. He worked long hours and retained the greater part of his salary for flying time. At least he was in the air again and on his way to a commercial license, for now he possessed the bare minimum requirements for acceptance as an airline co-pilot. He wrote to every known airline and requested applications for pilot employment.

The high point of his day was no longer the trip to the field, but rather the ride home to his apartment to check the mailbox. Day after day he received negative answers from the airlines. Their main complaint was his lack of flying time; he had only 400 hours.

Ed had a large check-off sheet he had made on a shirt cardboard. Here he kept accurate accounts of his applications. Hope rode with him. The word failure was not in his vocabulary. There were still four more airlines to hear from. But things were beginning to look pretty grim, and dejection set in. He began to contemplate getting a job as a banner tower or crop duster.

Finally a letter came from Eastern Airlines requesting that he contact their office in Manhattan. He was elated. At last he would have a chance to be an airline pilot. He passed the preliminary tests in New York, then was sent to LaGuardia Airport for a physical, and to meet the chief pilot.

The interview opened with, "I'm Charlie Pratt, have a seat."

"Thank you, sir," Ed replied, sitting on the edge of his chair, feet and legs together as though he were back in the cadets.

"Relax, you're not in the army now," said the chief pilot. "I see that most of your time is civil. What about your hitch in the Air Corps?"

Ed swallowed, cleared his throat, and wondered if he should tell the truth or hope the airline would never go into the matter any further. His friend, Charlie Root, had washed out of flying school. He told Pan American that he had never flown in the service. He was halfway through ground school when he was terminated for lying on the application. Ed made up his mind that truth would probably be the best thing.

"Sir, I had no problem flying the J-3 Cubs. It wasn't until I tried the BT-13 in basic school that I had the difficulty."

"What kind of difficulty?"

"Well, I just couldn't get the hang of the thing in the time allotted. My instructor said that they couldn't give me any more time. There was a war to fight and you know the rest of it."

"Well what makes you think Eastern can make a pilot out of you? You know that we will look you over as much and more as the military. Our passengers' safety is paramount."

Ed leaned forward. "Captain Pratt, I know my time is low and my experience is solely in light stuff, but I'm sure I can do the job. I have wanted to fly all my life and if you would just let me prove it to you I won't let you down."

Pratt scanned the application again. After a few minutes he put it down and said: "I certainly admire your honesty and persistence. I am going to take a chance with you and put you down for a class starting in two weeks."

Ed was delighted. He rushed over and shook Pratt's hand as though they were long-lost buddies.

"Thank you, captain," he beamed. "I really appreciate what you have done for me."

Quinn was not the cleverest fellow in the world, but at least he knew it. During the month-long ground school he studied continually. Most of his fellow students had been trained by the military, and many of them had flown four-engine equipment.

Each night Ed would return to his room and study—harder than he had ever studied. He wondered if he were being trained as a pilot or a mechanic. Diagrams of the electric, hydraulic fuel systems were in his mind constantly. He felt the training was good but why did they expect so much? After all, if he couldn't push a button or pull a lever in the cockpit to eliminate a malfunction, why study about it?

When the results of the final phase were posted, two of the fifteen students had failed. Ed was not one of them. He was thankful that he was still in the program, but was keenly aware that the flight checks could be his undoing.

He was scheduled to fly at 2 P.M. the next day. Since the airlines run a strict seniority program he would be trained on their lowest-paying and smallest plane, the Martin 404. A tricycle-geared forty-four-seat twin-engine monster. Captain Bob Stanfield would be his instructor, the man who could make or break his future.

Even if he passed the check, he would be on a year's probation. The Air Line Pilots Association will not take a member under its protective wings until he completes his first year. During that time each line captain that a first-year man flies with has to submit a written report to the chief pilot. This all-inclusive check sheet is concerned with appearance, ability, judgment. There is also a section wherein the captain expresses his comments or recommendations. More than one pilot has flown for eleven months only to be terminated before his probation is up.

Captain Stanfield was a respected senior captain with the line and had been in the training department for two years. He enjoyed the challenge of turning out good pilots, and he was well known as a man that would make the fledgling as relaxed as possible.

Before each flight a thorough outside inspection must be accomplished. This is called the walkaround. Even though each plane is preflighted by a mechanic, it is still the pilot in command's responsibility to check for himself.

As they both settled into their seats after the walkaround, Captain Stanfield scanned the logbook. Finding no discrepancies, he signed it, signifying to the CAA (Civil Aeronautics Administration) that he was accepting the plane for flight.

"Ed," Captain Stanfield said, "this machine is nothing more than a kiddie car with wings. I know that you have heard stories about the poor performance during warm weather. Since the air is less dense when it's warm, the props and wings don't have a good chunk of solid air to bite into. If you lose an engine on takeoff you had better move fast and be sure. Fly it every second and don't feather the wrong one. I'll run the check list and you sit there and respond to my instructions. I know this is the reverse of what you'll be doing on the line, but I want to acquaint you with the location of the switches."

Ed's heart started to pound. His hands grew clammy with each of the myriad switches he fingered. Engines were started, and when a break in the steady patter of radio instructions came, Ed picked up the mike.

"LaGuardia ground, Eastern trainer 22 ready to taxi," he called.

"Roger, Eastern trainer, taxi runway 31, altimeter 30.00, wind 300 degrees 10 knots, hold short of the outer."

Ed made the takeoff and did an average job of keeping the plane straight. Since the captain has the only ground-steering control, the first part of the takeoff run is up to the man in the left seat. The climbout was good, and as they turned toward the practice area Captain Stanfield hung up his earphones and suggested Ed do the same.

"We don't need to listen to all that chatter," Stanfield said.

"Thanks, I will."

After twenty minutes of turns, climbs, descents, Stanfield reached down into his flight bag and came out with a piece of aluminum that he clamped on to the windshield directly in front of Ed.

"There, that's better for you," he said. "Now the sun won't bother you."

The use of the aluminum is primarily to restrict the training pilot's vision so that he must rely solely on his instruments to keep right side up. Since there is no visual reference in clouds, this is the best way to simulate flying in weather. It is a well-known fact among pilots that weather flying separates the men from the boys. Some pilots can fly like eagles when they use the ground for reference, but when they have to fly by "the gauges" they are all over the sky.

Ed was no different from other beginners, except he let his altitude vary considerably. When he was told to check the altimeter his airspeed would suffer.

"Settle down, Ed," Stanfield warned. "Don't fight it. Treat her as though she were a girl you wanted to make points with. Be smooth and gentle. Watch me."

Almost immediately, as if by magic all the instruments stopped their dance, and settled down as though locked in place.

"You have it now. Relax." Stanfield reached into his pocket and pulled out a pack of cigarettes. "Care for one?"

"No thanks. I'm having enough trouble here without trying to smoke."

"Now you're getting the hang of it. Turn to 090 and climb to 8,500 feet."

Ed advanced the power and eased the wheel back. Almost immediately the rate of climb soared upward and the airspeed got dangerously low. He caught the error in time and relaxed the pressure on the wheel.

He knew things were not going well, and he grew more tense each minute. There was a decided yaw to the plane as they passed 7,000 feet, and a tremendous amount of right rudder pressure was required. A quick scan of the flight instruments showed a continued turn to the left, and a rapid loss of airspeed.

Something vital was wrong, but what? He glanced once at the check pilot, but got no reaction at all. He was calmly smoking his cigarette and looking out his side window. Ed was dumbfounded by Stanfield's nonchalance.

More power was advanced, and the airspeed was still low in spite of the downward trend of the machine.

"What seems to be the problem?" asked Bob Stanfield.

"I don't know. It sure feels funny. I can't maintain altitude or airspeed."

"Have you checked all the instruments?"

"Yes sir, I have."

"Well, are they all normal?"

The beads of sweat were edging their way out all over Ed's body, and he kept thinking: I have to do it right. This might be my last chance.

"What about the cylinder head temp on the left side. How does it look?"

"It's very low, sir."

"What would cause that?"

"I don't know, sir. All the other gauges seem normal." Ed was panicky.

Stanfield put his cigarette out and explained. "You've been on single engine for some time. I cut the fuel rather than pull the throttle. It's a lot more difficult to detect that way. The windmilling prop makes the gauges read near normal. Since the engine is being driven by the prop, the lack of work is first evidenced by a decided yaw and a cooling engine temp. You take it easy. I'll fly her back to the field."

Ed knew the ride was bad and they might not give him another chance. He reported to the director of flight operations and his suspicions were confirmed. He was told that lack of experience combined with a very poor check ride gave them no alternative but to let him go.

Dejected, his self-satisfaction dead, he went out and drank himself into a stupor. It took months before he could face the fact that he really wasn't good enough to fly for an airline.

He borrowed enough money to buy an old Fleet biplane and started a business towing advertising banners. He ran it for nearly a year and was just starting to break even when another chance at the much-cherished airline job opened up before him. Ed had just put in a long day towing banners up and down the south shore of Long Island with one or two trips over Manhasset Bay. He was about to close the door on his hangar when the office phone rang and he turned back to answer it.

An excited voice on the other end said, "That you, Ed? Remember your long-lost roommate from old 41G, John Barker?"

"Do I? How the hell are you?"

"Listen, Ed, I don't have much time to tell you, but are you still interested in an airline job?"

"You better believe it."

"OK, then. I have a little pull with the chief pilot, and I think I can get you in. I have been a captain on X Airlines for over five years and I occasionally give a few line checks. I'll drop you a note with all the poop. Must go now."

With this sudden rising of his hopes, he looked at himself in the mirror. His gaze met eyes that were steady, implacable, arrogant. Within a week Ed Quinn was interviewed by the chief pilot of X Airlines.

The preferred ex-military pilot applicants were running thin, and Ed's 1,000 single-engine hours were beginning to look good. The age-old story of supply and demand is no different for airlines than for any other business. And perhaps another cliché should be added: the right place at the right time.

Ed's timing was good and he was assigned to ground school starting in a month. He got through all phases of his training and was lucky enough to get Captain Barker on his check ride. Flying a DC-3 was considered less complicated than the Martin at Eastern.

His limited ability was overshadowed by his ego. Everybody on the line liked him, from the ramp agent up to the president. He went out of his way to butter everyone up. Most of the tower personnel called him by his first name and, time permitting, he would visit with the controllers and buy them coffee. Being low on ability and high on self-assurance, he daily sowed seeds he hoped would pay off when the time arrived. A cup of coffee and a friendly bit of chitchat now and then may keep a violation from blackening a record.

The DC-3 still remained a challenge to him. That's the kind of bird that can be kissed onto the runway time and time again. And just when you start to feel your mastery over her, she'll suddenly head for all parts of the strip, except the end. Most pilots can relax during the landing rollout; not so the pilot of a DC-3. When the tail hits the ground, that's the time to be wary.

The three main steps in a pilot's career are (1) to get the job, (2) to survive the first probationary year, and (3) to check out as captain. Ed had the first two completed. Only one more to go, the biggest of all.

To compensate for his lack of ability, Ed got well established as a company man with a capital *C*. Every pilot will help his company if there is no safety factor involved, or union or government rule to bend, but First Officer Ed Quinn knew no bounds. He was known as good old Ed, the man who would take any trip, anytime, anywhere he was told. Not a safe way to operate, but at X Airlines it was the greatest way a pilot could make points with the company.

New routes were awarded the airline and Ed could see that soon it would be time to upgrade to the left seat. The next crew alignment published showed Ed Quinn as captain on the DC-3.

The required time to check out as captain was eight hours of intensified training before the CAA would ride with you to issue the coveted Airline Transport rating. Ed had double that amount of instruction. His "company first" attitude was now being repaid. Many safety-conscious pilots would have long since been eliminated from the program. After nineteen hours of instruction Ed Quinn was put up for, and passed, his rating ride. He was now a member of the elite who wore four stripes on their sleeves and scrambled eggs on their hats. Captain Quinn was indeed in command, but would he get the respect that the position demanded?

As first officer there was little harm he could do, other than fly when he should be on legal rest. Now as a captain the full impact of his subservience would be felt. If it weren't for X Airlines Captain Ed would still be towing banners—a thought that was ever present in his mind.

Captain Quinn's unsafe tactics soon became common knowledge among his fellow pilots, and he became the butt of many jokes. Nothing short of a bent wing or a flat tire would keep Ed from taking his scheduled trip. His lack of judgment coupled with his pro-company feelings often caused him to brush with disaster.

Cleveland Municipal Airport had been blanketed with snow for two days. As snowfall diminished, a strong wind from the northwest developed, with gusts to 35 knots. The main northeast instrument runway was plowed full length and width, with a layer of ice and snow still present. No flights had landed since the snow began. Now the field was open, but because of the strong crosswind, no flights were operating.

Captain Quinn was en route from Chicago to Cleveland and was advised of the weather and runway conditions over the company radio. He switched frequency to Cleveland tower.

"Cleveland, this is X23, would you say your winds."

"Roger, X23, surface winds 300 to 320 degrees at 25, gusts to 35, breaking action poor on 5R, advise intentions."

"OK. Stand by."

A quick mental calculation showed a crosswind component

of over 35 knots, far in excess of the allowable limits, and with a slick runway. Only a fool would try it.

Ed turned to his co-pilot who was busy with graphs and regulations and whatever else he could find to keep the inevitable from happening.

"What do you think, Tom?"

"That's a pretty strong wind and no one else has tried it yet."

Ed thought for a moment, only to appear as though he were considering Tom's comment. "Tell them we'll give it a try."

As a new co-pilot Tom most certainly did not want to argue with his captain, but he knew the dangers involved.

"Cleveland tower, X23 would like to land."

"Roger, you're cleared to land 5, right wind remains the same and I'll give you wind checks on final."

"Roger."

There were four other planes circling the field awaiting wind change. On hearing X Airlines transmission they couldn't wait to add their comments.

"Good luck, you'll need it," said one.

And another voice came, "Better you than me."

Captain Quinn ignored the remarks and continued with his single-channel thoughts. He was being paid to take people from A to B and that he was going to do. Safety was a dirty word. They were now half a mile out on final approach with Ed's check list complete. He didn't believe in using the entire list so long as the wheels were down. That's all that counted.

Tower advised: "Wind is now steady 300 at 30 knots."

Ed was working feverishly to keep from drifting off the runway. He touched down within the first 500 feet, and pinned her on by applying forward pressure on the wheel. The tail was starting to settle and it seemed as though he had lucked out again.

Suddenly the realization that the landing was not over filled him with horror. Like a headstrong horse the plane swerved sharply into the wind and slammed into a snowbank. Both props came to a shuddering halt as the right landing gear collapsed. There was no fire and only a few minor injuries.

At the hearing that followed, the CAA awarded Captain Ed Quinn three months on the beach for careless and reckless operation. The company showed that it condoned Ed's spirit; to eliminate some of the penalties, they paid him for all vacation and sick leave he had earned so far. This would secretly eliminate one month's fine.

When Ed returned to work none of his arrogance was lost. In fact, as if to overcompensate, he acted as though he had never made a bad decision in his life. The fact that his fellow captains made jokes about him and wouldn't allow their families to fly on his plane appeared to have little effect on Ed's ego. All that mattered was pleasing the company. They paid him.

Captain Quinn's hair-raising exploits continued and he became the captain that first officers least liked to fly with. Trips were bid on a seniority basis. Since Ed was a captain that copilots tried to avoid, only the most junior men were compelled to fly with him. The only way that Ed could entice a senior man was by allowing him to sit in the left seat. This was strictly against company regulations. In order to remain undetected, Captain Quinn would taxi away from the gate and change seats when away from the watchful eyes of the company. On landing he would follow the same unorthodox procedure. He knew that if he were caught, little would be done to him.

X Airlines purchased a number of Martin 404 aircraft from TWA, who were converting to larger equipment. Since no one on X Airlines had any experience in these machines, TWA sold them a package of flight training along with the equipment. Captain Quinn bid and was awarded upgrade training at TWA's Kansas City complex. Here he would be just another pilot and his company-prone exploits would have no bearing on the TWA check pilots.

With years of command time under his belt Ed felt no qualms about flying the plane he had previously trained on with Eastern. He was sure that, even if he couldn't handle her before as a co-pilot, he would have little trouble now.

On his first training ride the check pilot cut one of the engines before V1 (after this speed you are committed for takeoff). Ed became confused and kept full takeoff power on the good engine. In trying to get into the air he nearly ran off the side of the runway. The check pilot took over and taxied to the terminal.

Captain Quinn was returned to X Airlines with another of many down grades on his record. After getting a recheck on the DC-3 he was permitted to fly as captain again.

Within six months, a number of X Airlines management pilots were qualified on the Martin. Captain Ed Quinn was permitted to try the Martin again. He failed. After numerous extra hours of instruction, he somehow managed to get his type rating.

X Airlines sends its weakest pilots to train with those check airmen who will give every consideration to the fact that the men they are teaching are pro-company. X Airlines has one supervisory pilot who also is authorized by the CAA to issue ratings. Most of Captain Quinn's proficiency checks were given by this man, and always at a time when the regular CAA examiner was off duty. No regular line pilot would ever pass a man who wasn't qualified.

Through the years, Quinn's obsequiousness persisted. He remained ingratiating to anyone in the company who might someday come to his aid. Eventually he found himself commanding one of the latest jet additions to his airline's fleet, a Douglas DC-9.

The Air Line Pilots Association is far more safety conscious than the FAA, and has definite on-duty limitations. The government has no regard for the amount of rest a pilot gets; luckily, ALPA does. They state that no pilot can be ordered for duty at his home station without having at least ten hours' rest.

Through poor planning, some pilots run out of flight time before the month is over. If the line cannot find men to man their machines, they are forced to cancel some trips. The FAA limits a pilot to a maximum of 100 hours per month. This may have been safe in 1935, but that same number of hours is in effect today. ALPA restricts the total monthly flying to 85 hours.

By January 30, and with a month of bad weather behind them, most of X Airlines' pilots had completed their time for the month. In spite of the fact that Quinn would not have his legal rest, and would also exceed his allowable flight time for the month, he was asked to take, and he accepted, a DC-9 flight.

It was a cold and snowy night when Quinn reported to LaGuardia Operations. Through the years his sight had slowly deteriorated and he now wore glasses. The cigarettes that he smoked continually left a trail of ashes down his unpressed uniform. He was a good example of the way a captain should not look. He casually glanced through the weather reports while the Operations clerk stood by.

"Hey, Bobbie," he called, "when do you expect the equipment in, that makes up Flight No. 23?"

"In about twenty minutes."

"Will we be on sked?"

"Probably a little late. Inbound flights are all experiencing ATC delays."

"OK. Tell my first officer I'll be in the snack bar. Check on the 'stews,' make sure they're here."

"Sure thing, cap."

During Captain Quinn's absence his first officer, Frank Beni, reported in and started working on the flight plan. It was a Friday night and there would be a full load of passengers. Dispatch release showed 24,550 pounds of fuel and Green Bay as their alternate for Milwaukee.

The two stewardesses for the trip were roommates. They arrived together, but because of traffic they were twenty minutes late. They were informed that the equipment that made up their flight would be at Gate 33. Since it would be a quick turnaround, they hurried out to ready the plane for their passengers.

Captain Quinn returned to operations and spoke to his copilot. "Hi, Frank. I see they caught you for this one, too."

"Well, I can use that time. I've had bad luck with cancellations this month."

Ed slid the flight plan over to him and signed it. "I'll get a full month," he said with a smile, "ninety hours or so. I'm going out to the machine. See you on board."

"OK, Ed. I'll be out with all the goodies."

As Ed walked toward his aircraft he met the incoming captain, who greeted him with a worried look.

"Hello, Ed."

"How's it going, Dick?"

"Well, the plane's OK. The only squawk I had was a high fuel flow on the right one."

"How high?"

"Well, at cruise she was showing 4,500 pounds."

"That's really up there. Do you think it was an indication problem?"

"No, I don't. Because the fuel gauge on that side was low also. It's all in the book so you can read it and weep. I have to get going. It's my anniversary, what's left of it."

"Thanks for the info. Have a good time."

"Thanks. See you."

As Ed entered the cockpit he saw a mechanic reading the log discrepancies.

"Hello, cap."

"Hi."

"Say, cap, this squawk might really hold up the departure. I'll give the engine a check, but if what I read here is true, we have a real problem."

"Well, I wouldn't get too excited over it. That last captain writes up ash trays when they're full."

"I'll unbutton it and see what I can find."

The wet snow was growing more intense and the men were having a tough time deicing. They had used so much fluid on the previous flights that their supply was almost depleted.

Captain Quinn was one of the mechanics' favorites. He seldom wrote up a discrepancy. The fewer squawks in the logbook, the more time the mechanics had in the coffee shop, so the more Ed would be liked in their eyes.

First Officer Beni settled into his seat. "She's OK outside," he said, "excepting the wings. They've cleared the biggest part of the snow off but there's an icy layer beneath that they can't penetrate."

It is against federal regulations to start a flight with any snow, frost, or ice adhering to the wings or tail. The shape of the wing can have the lifting qualities destroyed even with a thin layer of frost.

Before Ed could answer, the mechanic entered the flight deck brushing the snow from his face. "We found a leak," he said, "around the fuel controller. I tightened it up. But it's up to you. We're also out of deicer fluid. If I go back to the hangar it will take twenty minutes to fill up."

By now the gate agent was getting into the act. "Say, captain," he asked, "is it OK to board now or will there be a delay?"

Ed knew better than to fly a trip with such glaring discrepancies. Still, eighty-eight people were waiting to board—that meant a lot of money to the company. His mind was made up.

"Everything is OK. Button up the engine and load up."

The passengers hunched as they waited in line in the snow. They resembled turtles who did not like the weather and had withdrawn their heads into the safety of their shells.

The manifest was brought forward by one of the stewardesses. As expected, it was a full boat. A quick look by Frank at the takeoff charts showed the decision speed V1 of 136 knots; V2, the minimum climb speed, would be 141 knots with 15 degrees of flaps.

The engines were started and the before-taxi check was read by Frank and answered by the captain.

"Electric panel?"

"Checked."

"Galley power?"

"On."

"Seat belts? No smoke?"

"On."

"Pilot static heat?"

"On."

"Boost pumps on?"

"Center off. Mains on."

"Air conditioning supply?"

"Auto."

"Your damper?"

"On."

"Load lights?"

"Out."

"Hydro pumps?"

"Four on system normal."

And so on and so on.

Captain Quinn responded to the waveoff and eased the throttle from idle to 85 percent. It took considerable power to get the heavily loaded giant away from the gate. The ceiling was now 800 feet, visibility 2 miles.

"La Guardia ground, X Airlines 23 IFR to Milwaukee."

"Roger, 23. Do you have your clearance yet?"

"Negative. Unable to contact clearance delivery."

"Stay with me. I'll get it. Clear to runway 13, hold short of 4. Wind is 100 degrees at 18, gusts to 22."

Flight 23 was No. 3 for takeoff. Ed flashed the wing inspection lights on to check the snow buildup. Frank was afraid to look; he knew the snow that would probably blow away on takeoff was only a covering for the ice beneath it.

La Guardia tower finally came through with the takeoff clearance and Flight 23 taxied into position. The final portion of the check list was completed and the tempo of the windshield wipers was increased.

"All set," said Quinn.

"Roger," replied the first officer.

As Ed advanced the power the low whine of the engines increased their intensity. Instantly the lazy instruments sprang to attention and raced to the green marking on their cases as 105,000 pounds of dead weight started to move.

First Officer Beni was guarding the throttles and calling out specific information. "Engine instruments in the green epers OK overhead, panel good. Eight knots."

At this speed the captain relinquished the nosewheel steering and placed his hands on the flying wheel. Now there would be sufficient airstream over the rudder so that it could be used for directional control.

The airspeed slowly crept past the 100-knot mark and only a few thousand feet of runway remained. At 136 knots Frank called "V1."

In what seemed an eternity the plane finally climbed to "take-off" speed. Frank yelled "Rotate." Simultaneously the red obstruction lights on the dike at the runway's end loomed out of the snow. Quinn applied normal back pressure to the wheel but got no reaction. She was heavy all right but she should have been in the air by now. He applied more pressure, and as Frank called "V2" the plane staggered into the air only inches from the dike and certain death.

Ed called "Gear up." And the powerful hydraulic pumps started retracting the wheels into their resting places. The three gear-position lights extinguished and the climb check list was called for. A right turn was started for noise abatement. A turn that directs the plane to Shea Stadium.

The normal cockpit noises suddenly were forgotten when the spine-tingling bell of the engine fire-warning system sounded, accompanied by three red lights.

"No. 2's on fire!" cried Beni.

"Shut it down and fire the bottle. Tell them we're coming back."

The altimeter started to unwind and the airspeed dropped. Full power was applied to the good engine which temporarily cut the descent. The craft was shaking, the advance warning of an impending stall. Now Ed was wishing he had heeded his co-pilot's warning about the ice on the wings.

Captain Quinn was caught in his own trap. Up until now he had been lucky. The lights atop Shea Stadium grew larger by the second. To Ed it resembled a giant mausoleum, a potential resting place for Flight 23.

There's one way to aggravate a stall and that's to turn. Ed's plane was teetering between flying and crashing. It was obvious that the precious altitude could not be traded for a turn. He elected to hold a straight course and pray. The altimeters were watched desperately, their importance being second only to the airspeed which, if allowed to drop, would certainly give the newspapers a devastating headline.

Slowly, foot by foot, the long needle of the altimeter began to climb. Ed's throat was dry. His veins resembled ropes as the blood pumped rapidly through them. He was now more like a nervous sparrow than the fearless eagle he portrayed on the ground.

Miraculously, Flight 23 skimmed over Shea Stadium. There still remained the task of getting back to the field, however. The altimeter was now indicating 400 feet and increasing slowly. A gentle left turn was started and a few feet of altitude were lost.

Back at LaGuardia, all crash trucks were alerted and in position at strategic points along the edge of the runway. Traffic to the northeast of LaGuardia was diverted to allow a clear path for Flight 23.

Quinn had managed to negotiate a turn to the northeast and was on a wide downwind leg for runway 13. As he passed over the Whitestone Bridge he was below the obstruction lights of the towers on each end. The signs on the trucks that were jamming the bridge below could be read with ease. The good engine was starting to overheat and the throttle was reduced slightly. Immediately the airspeed approached the danger level. Power was promptly increased before too much of the precious altitude was lost.

Frank was in contact with approach control and the radar man was issuing steers to the field. Only one major obstacle stood in the path of Flight 23, the huge natural gas tank that rose defiantly 422 feet tall at the south end of the Bronx. For years pilots had complained about that tank as a hazard to flight.

A new flame of fear leaped up in Quinn's mind. Will I be the one to go down in the annals of aviation as the man who wiped the gas tank out?

Radar called: "Flight 23, I can take you south of the tank if you like."

"Negative," replied Ed. "I can't make a turn that sharp. I have it in sight so far. I'm going around the north side."

"OK, sir, we'll keep watching you. You are clear to land. Men and equipment are standing by, lights are on step 5."

"Leave them up until I tell you."

The tank appeared ahead to the left. Flight 23 would never have topped it. Quinn could see the lights on the five stacks of the power plant. He started a gentle turn and picked up the runway lights. Now it would be all downhill. He waited until he was two miles out before dropping the landing gear. Three green lights flashed on signifying the wheels were in position.

The check list was completed and the bug speed was set at 135 to allow for the overweight landing. There would be no room for error on this approach, because to go around and try again would not be possible.

By the sheerest luck, a disaster was averted—a disaster that should never have arisen in the first place. Here are some of the errors made by Quinn:

1. Flying without sufficient rest
2. Exceeding his maximum time for the month
3. Departure with ice on his aircraft
4. Insufficient inspection of fuel leak that caused a fire
5. Making dangerous turns to return to LaGuardia when he could have landed straight ahead at JFK

With a few lies here and there, Quinn managed to be commended for his actions instead of being disciplined. Most people learn by experience, but not Captain Lucky. He is as dangerous today as he ever was.

Chapter Six

OUTER MARKER, INBOUND

ON NOVEMBER 3, 1962, at 9:30 P.M. the following weather information was broadcast by Idlewild Approach control.

> All aircraft copy. Runway 4 Right I.L.S. in use, Landing 4 Right. Wind northeast six, altimeter 30.31. Idlewild weather, sky partially obscured, visibility 1½ miles with ground fog, and, just got out new visibility, it's one mile now. Runway visual range 4 Right inoperative. Middle Locater 4 Right inoperative and precision approach radar not available.

Eastern Air Lines Flight 512, Captain Edward J. Bechtold in command, was being laddered down in the Sandy Hook holding pattern. It was presently No. 3 for landing. Captain Bechtold would soon be vectored to the ILS approach course and down the electronic beams to a final landing and home.

At 9:41 a Sabena jet located on the ground one mile from the active runway was cleared for takeoff on runway 7R. The Sabena crew informed the tower: "The visibility is a bit too poor, we request to hold."

At 9:43 EAL contacted the tower and advised: "Outer Marker, inbound." (The final approach fix, approximately five miles from the airport.)

Tower replied: "EAL 512 cleared to land."

Now starts the most critical part of an instrument approach. In a matter of seconds a pilot must decide if he can make the transition from instrument guidance in the cockpit to visual ground reference.

As Captain Bechtold approached the roundout altitude over the runway, his contact with the field vanished and he was immediately engulfed in ground fog.

He gave the order to go around, and as he applied the power Nos. 1 and 2 engines struck the ground and were ripped from their mounts. The nose section contacted a four-foot mound of dirt, causing the aircraft to ricochet back into the air, passing a taxiway, and then dived savagely back to earth. The right wing parted and burst into flames. The crash killed the crew and twenty-two of the forty-eight passengers on board.

It took twenty-nine minutes for the rescue crews to get to the wreckage. Were it not for the airport detection radar (ASDE radar so accurate it can spot a man miles away) it would have taken longer. The contributing factors to the accident follow:

1. Inadequate weather information
2. RVR instrument inoperative
3. Middle marker inoperative (a radio transmitter that activates a flashing light in the cockpit usually 3,000 feet from the runway threshold)
4. No precision approach radar (the most accurate guidance to a landing affording both direction and altitude)
5. Surface of the ground adjacent to runway exceedingly rough

The primary reason for this disaster was the rough ground between the runways. It literally tore the aircraft to pieces; the landing could most likely have been survivable for all.

Chapter Seven

THE BEGINNING AND THE END

THE AIRPORT is the key to traffic delays. Not only are there not nearly enough airports, but the majority of them are inferior. Compared with today's jets, the landing sites belong in the horse-and-buggy era.

Airport planners can learn a lesson from the railroads. Trains have relatively few rails between points A and B, but branch out to many tracks to facilitate the numerous trains at a main station. Airplanes are flown over many different routes between A and B. But when it comes to airports, the aircraft are strung out for miles all headed for the one lone strip.

Washington National has a modified ILS approach to the south. The whole idea is to keep planes from flying over the Capitol so there is an ILS that leads to nowhere. By that I mean it doesn't align the pilot with the runway, but it does keep him over the river. When the plane breaks through the clouds it is necessary to change heading 34 degrees in order to land. Not a very safe maneuver so close to the ground.

Kennedy has an approach that is the granddaddy of the

Mickey Mouse variety: it's called Canarsie VOR. This approach is made from the ocean toward Canarsie. When the pilot overheads the nav-aid, he must strain to find the lead-in lights that are supposed to take him to the runway. Then the pilot has to make a 90 degree change in heading to land. The whole approach is useless, and as for lights, the Belt Parkway has lights that are easier to see. Many a pilot has mistakenly followed the highway rather than the lead-in lights. There is a perfectly good instrument-landing system on this runway but it is rarely used because the aircraft would be making too much noise over Brooklyn. The total cost of installing the Canarsie approach was over $250,000, nearly double the price of an ILS.

When JFK Airport was built, the FAA estimated it would be able to handle a maximum of 260,000 aircraft movements per year without delays, yet it has already exceeded the 500,000 mark. The number of airline passengers will increase to 269 million in 1974 and will be up to 429 million by the year 1979. We are already behind the power curve, and will have all we can do to keep above water, let alone expand to a comfortable level.

Senator Mike Monroney, in a speech before the Air Line Pilots Association in June of 1967 said, "We are being forced to handle the 'Airport Problem' on a crisis basis. That's not the best way to solve any problem, and certainly, it will be a much more expensive one than if we had provided enough money five years ago to fund a realistic improvement program.

"I predict that it will be many, many times more costly in 1977 to both the government and private industry, if we don't authorize and fund a sound program for airport and air traffic control improvement now."

During the past ten years, federal aid to airports has averaged $28.5 million per year. This is approximately 10 percent of the total cost to build Dulles Airport.

In fiscal 1970 $4.58 billion was allocated in the federal aid-to-highways program. In fact, 71,336 miles of single-lane highways are constructed each year, but it requires a serious crash to secure an extra 1,000 feet of life-saving runway.

Not only is a federal aid-to-highways grant easy to get, but the government pays 90 percent of the total project cost. For airports, 50 percent is the allowable maximum.

ALPA believes that airports and their managers should be certified, as are other segments in aviation. The aircraft pilots, mechanics, controllers, dispatchers, and weather observers are all licensed. Why not the airport? Our inadequate airport and airways system, with its continual lack of landing aids and airport safety standards, has caused substantial loss of life, delays, and adversely affected the growth of aviation.

Lives and aircraft have been lost because of the following:

1. Obstruction hazards near the airport
2. Lack of underrun and overrun (additional paving on each end of the landing strip)
3. Noncompacted shoulders adjoining runways (even highways have this)
4. Lack of sufficient runway length and number
5. Lack of level land adjacent to runways

Commercial air carrier statistics for the years 1957 to 1967 show a total of 138 accidents: 28 overruns, 46 underruns, and 64 veers. One DC-8 landed short of the runway by 18 feet, wiping out the landing gear and thoroughly scaring 98 passengers. A 707 landed long and crashed into the ILS shack at the end of the runway, closing the field for 14 hours: it needed 425 feet more pavement.

There are a number of reasons why an aircraft will not stay where it is put, or stop when asked: among them are wet slippery rubber deposits, smooth-tread tires, crosswind, and hydroplaning (a thin film of water on the runways can actually raise the tires so they no longer contact the surface, which makes braking impossible).

On July 1, 1965, a Continental 707 with sixty-four people on board touched down at Kansas City's Municipal Airport. After making a normal ILS to runway 18, ground contact was approximately 1,050 feet from the approach end. (This is where the glide slope usually intersects the runway.) Reverse power, boards and brakes were properly applied, and utilized for the remainder of the 5,950 feet of hard surface. White tire marks characteristic of hydroplaning were evident from touchdown until the craft left the pavement and slid off the runway into the mud. The fuselage broke into three parts, making the $5 million machine a total loss. Luckily only a few people were injured.

Logan International Airport in Boston, Massachusetts, has its main instrument runway shortened by 3,410 feet, leaving a marginal 6,590 feet for landing and rollout. This restriction is in effect because the approach path is too close to the Navy Yard. When the tower advises that vessels are maneuvering in this area, the landing limits jump a quarter of a mile. once a year an aircraft skids off the far end of this strip and bogs down in the mud of Winthrop Bay.

The longest runway at Chicago's Midway is shortened by 500 feet to allow for approach obstructions. This makes the strip 500 feet short of the ALPA minimum criteria for jet operation.

The crash of a Trans Caribbean Boeing 727 in the Virgin Islands in December 1970 was partly caused by a marginal runway. ALPA believes that a minimum of 6,500 feet is needed for a safe jet operation. Yet the total usable length of the St. Thomas runway is 4,650 feet.

Not one runway at New York's LaGuardia Airport has any overrun area to allow for error. Most of the airport is submerged when a heavy rain hits it. There is a dike around the field to keep the bay from spilling into it. If a pilot lands short, he will most assuredly either rip his wheels off on the dike, or on a pier, or perhaps land on the parkway. LaGuardia is not the worst culprit, however. The airport in Washington, D.C., has as its longest runway one that is 130 feet shorter than LaGuardia's.

Syracuse Airport in upstate New York has a runway with approach and center-line lights, touchdown lighting and overrun, an airport that caters to a fraction of the airline traffic of

LaGuardia or Washington. Yet the runway is 2,000 feet longer than either of the busy terminals.

In the summer of 1965 the FAA issued a new regulation requiring a 15 percent increase to the effective runway length (the total usable portion of the runway, allowing for safe margin of altitude over the approach obstructions) for all turbojet aircraft landing on a wet runway. This law means that a trip going to a wet field will have its total weight restricted, a good safety factor, but only a substitute for a lack of concrete.

A salient factor when landing on a short runway is the fact that the pilot knows he must touch down fairly close to the threshold. Here he runs a chance of landing short. If the field is icy or wet, the undershoot tendency becomes more acute.

If a landing strip slopes, up or down, it can affect the pilot's judgment. When either the terrain at the approach end of the strip or the runway itself slope up a pilot can expect an "above glide path" illusion. (He will be lower than he thinks he is.) The opposite is in effect for a downsloping approach zone. Correcting the illusion on the supposedly high approach results in landing short, and vice versa.

Problems in depth perception occur when runway color blends into the terrain. The concrete runway surrounded by sand or the asphalt strip surrounded by dark foliage cause similar problems. These items are never considered when an airport is built.

Two sure ways to eliminate the tricks a pilot's eyes may play are to have adequate landing aids, such as ILS, or a visual approach slope indicator (VASI), which consists of a series of colored light bars that tell a pilot when he is in the proper approach path.

Obstruction criteria have long been a sore point around airports. In 1950 the aviation industry became aware of community encroachment. Builders erected homes, factories, and theaters almost on the edge of runways, leaving no room for airport expansion, and giving cause for residents to complain about aircraft noise. General James H. Doolittle headed a committee to study airport needs. He recommended that each runway approach have a clear zone at least a mile in length and 1,000 feet wide. This area should be clear of houses or any other obstructions, and considered part of the airport. A few airports were built to these specifications, but it wasn't long before the dollar became more important than safety and some of the land was sold.

At Miami International Airport a hotel was recently erected at the middle marker. This structure is directly in line with the runway and extends 65 feet into the approach area. It is only 2.5 feet short of exceeding the present criteria, and when planes make their approach they will miss the hotel by a mere 115 feet, certainly leaving little room for turbulence or error. Miami Airport is just one of many that has dangerous approach conditions.

Since it seems impossible to get runways long or wide enough, one excellent means of upgrading the existing runway is to groove it. One-quarter-inch diagonal cuts are made in the runway. This increases the coefficient of friction to a point that a wet runway can almost be classed as dry. Another, more costly method of reducing skidding is by the application of a special epoxy that can be sprayed or troweled on.

Military airfields have for years used arresting devices at the ends of their runways. Although these differ somewhat, they all produce the same effect: keeping the aircraft from going off the far end of the strip. They are usually activated by the tower. By throwing a switch, an arresting net is positioned across the runway. As the aircraft wheels engage it, they pull on some extremely heavy chains which cause rapid deceleration. Flight Safety Foundation, Washington, D.C., ran a survey concerning the possibility of these arresting devices for air carriers. During the years 1960–67 the U.S. Navy had a remarkable number of successful arrests. These saves averaged 2,000 with a 99 percent reliability. Air Force records showed that 97 percent of all 1957–67 arrests were excellent.

The next eleven years have a potential of seventy-one accidents that could be prevented by runway arresting systems. Projection of the overall losses resulting from jet aircraft accidents forecast that these accidents will result in 300 serious injuries, 12 fatalities, and a monetary loss of $363.6 million. It is conservatively estimated that it would cost a million dollars to equip both ends of a runway with this system. At that price it would be possible to equip 360 runways with the potential savings from all forecasted accidents. The latest jets are over five times more costly than the 707s. The Boeing Company estimates the replacement cost of its 747 at $25 million, and the SST cost $40 million to replace.

In Denver, Colorado, on July 11, 1961, a United Air Lines DC-8 had hydraulic problems on the runway. Reverse on Nos. 1 and 2 engines was selected and failed. The aircraft yawed and left the runway, clipping a maintenance truck. The fire that followed the collision with the vehicle fatally burned 17 of the 122 persons on board. But for the panel truck, little or no damage would have resulted.

Here is another example of inefficiency. On November 23, 1964, TWA Flight 800, a Boeing 707, slid off the runway at Rome, Italy, and killed fifty-one people. Construction was in progress at the airport, and the already marginal runway of 8,612 feet was shortened by 2,050 feet. A road roller was working 100 feet from the active runway. Shortly after takeoff thrust was applied, a reverse warning light came on in No. 2 engine. The craft became impossible to control, and as it left the side of the runway it clipped a piece of grading equipment and burst into flame. This accident was 100 percent survivable. The main culprit in this story was a steamroller. Construction work at airports should be done at times when the least aircraft movement is expected. And by no means in close proximity to the active runway.

The *United Air Lines Cockpit*, a safety supplement, had this to say concerning runway hazards.

> Obstacles surrounding the sides and ends of runways were the major contributing factor in take-off and landing accidents of commercial aircraft in which fire was involved. It is estimated that 79% of the fire fatalities could have been avoided if the runway and overshoot areas would have been clear of obstacles.

Only at military fields do you have the safety of knowing just how much of the runway is left. They have numbered signs spaced 1,000 feet apart at the edge of each runway. In this way a decision to continue or abort the takeoff is made considerably easier.

I have flown into many airports all over the world, and few have a clean bill of safety. I know of no airports that have parallel taxiways far enough away from the active runway. Any plane that is experiencing hydraulic trouble which would adversely affect the steering or brakes is a potential hazard. Given the right set of circumstances, a faulty aircraft could wipe out two or three planes waiting in line. No thought is given to this by the tower operator who usually tries to get as many aircraft as close as possible to the active runway.

On September 8, 1970, a Trans International Airlines DC-8 crashed on takeoff at JFK Airport in New York. Eleven people died. The NTSB attributed the probable cause to an asphalt-covered stone approximately 1.5 inches in diameter. Since construction is nearly always in progress at many airports, the flights continue in spite of it all. As this DC-8 was taxiing in the vicinity of fresh asphalt, the wake of the engines hurled a stone up on top of the tail. It rolled back and lodged between the horizontal stabilizer and the elevator. (The elevator controls the up and down movement of the craft.) As the pilot applied up-elevator pressure on his controls the stone jammed them in this position and the aircraft stalled and crashed.

A $10 million aircraft and eleven lives were lost because of a stone. The NTSB requested that the FAA alert all DC-8 operators of the elevator-jamming potential. But no mention was made of keeping aircraft well clear of construction areas with their assorted steamrollers, bulldozers, and cranes that have snuffed out the lives of more than one air traveler.

Most hub airports have extremely complicated taxi patterns, and almost no signs show the name of the taxiways. Only a handful of the major airports have maps of their taxiways depicted in the pilot's manuals.

The control towers at many busy airports do not always have a complete view of the airport. Many fields such as Atlanta and LaGuardia have buildings that obscure the view from the tower. Nearly all of the vehicular traffic at airports lack radios, yet the tower is supposedly in charge of this traffic. Busy airports today look more like training grounds for demolition stockcar races than they look like airports. At least these airports have towers, and that is more than can be said of over fifty airports served by scheduled airlines.

If the FAA really cared about safety they would put towers and VASI approach lights at all airline airports. Instead, they base their sole criteria on the number of flights per year that each field records. An airline airport is eligible for a tower when it records 24,000 or more annual operations. For a nonairline field, this number is doubled. To get the VASI, an airport must have 5,000 landings per year on the runway.

How does a pilot know when it's clear to land at one of the smaller airports? He takes a good look, crosses his fingers, and hopes for the best. These small fields abound with student traffic. Aircraft as large as 727s make daily scheduled trips into these death traps. In 1968, 719 near misses occurred in terminal areas, most of these around airports without control towers.

The pertinent weather, wind, altimeter setting, condition of the runway, and traffic are issued to the pilot by the same fellow that checks your baggage. He is sent through a hurry-up course in weather, and after three days is issued a ticket that enables him to issue weather to airlines. Weather reports at these stations are usually erroneous, they just want to get the flight in.

Runway conditions at the hub airports are taken by men with as little experience as the bag smashers have at local airports.

To a captain, the braking action of a 747 is vitally important. He is controlling the destiny of 564,000 pounds of metal and 374 passengers. Yet the runway condition is reported by the driver of the snowplow. When a pilot lands and gives the braking action as poor to nil, the airport is in danger of closing down. No aircraft are permitted to utilize a runway that has nil braking action (i.e., stopping is impossible).

Since this puts an economic burden on the airport and its various facilities: fuel, restaurants, limousine services, connecting flights, gift shops, and last but not least the authority that runs the airport itself, another check is always made. Now the county cop or Port Authority is told to get out and make a runway check. He usually does this in a jeep or station wagon equipped with chains and snow tires. After getting up enough speed to make the sham look good, he slams on the brakes. He may call the braking good or poor; but you can bank on the fact it will never bc nil. Of all the braking reports I have ever received, about one-quarter of them were accurate.

Every aircraft that lands at a commercial airport has to pay a fee. It is based on the weight of the plane. At JFK the minimum fee is $25 for a light plane. Airlines pay 35 cents per 1,000 pounds of gross weight. Every Boeing 747 landing costs its company $197.40. The annual landing fees paid by one small airline to use the runways at Pittsburgh International are $877,000. Total U.S. landing fees exceeded $150 million per year in 1971. Thus, in order to keep the money rolling into the pockets of the airport authority, airports are seldom closed.

Many airline passengers who survive the initial impact of a crash are later burned to death needlessly. Airports as busy as Palm Springs, California, Panama City, Florida, and Rochester, Minnesota, have no fire-fighting equipment on the field. These airports handle scheduled four-engine jets carrying over 100 passengers. Small-airport passengers have as much right to live as do those who land at Los Angeles International.

What happens when a plane's landing gear folds up and a fire breaks out? The tower, if there is one, summons the closest fire department. They may arrive in time to help remove the bodies. In most cases, off-airport equipment is for structural use and not suited for crash fire-fighting.

The March 1969 ALPA survey of Airport Fire and Rescue Facilities listed 22 airports served by twin-engine jets that had no fire-fighting equipment. In addition, 167 fields used by scheduled piston and turboprop airlines were in the same boat. The so-called safety-conscious FAA could and should make adequate fire-fighting equipment mandatory at all airports.

ASDE (Airport Surface Detection Equipment) is vital equipment at all hub terminals. When the weather is bad and visibility is cut to less than a mile, this sensitive radar is used to guide planes on the ground. Many towers are a mile or more from the active runway and they cannot tell a plane's position if they cannot see the plane. With this fine instrument, collisions can be avoided and rescue equipment guided to where it is needed. In spite of the necessity for this radar, it has slowly been eliminated at major airports. In 1964, eleven sets were in operation, today there are only six: at San Francisco, Washington, D.C., Portland, Tacoma, Newark, and JFK.

ALPA strongly believes, and can prove, that during wet weather a turbojet (not prop-driven) needs at least 6,500 feet of runway for a safe operation. Over fifty airline airports do not meet this criterion, among them: Bakersfield and Santa Barbara, California; New Haven, Connecticut; Pensacola and Daytona Beach, Florida; Kauai, Hawaii; Baton Rouge, Louisiana; Erie, Pennsylvania; Wheeling, West Virginia; Rochester, Minnesota; Albany and Binghamton, New York; Akron, Ohio; Norfolk and Roanoke, Virginia; St. Thomas, Virgin Islands. JFK's longest runway is 14,572 feet, all of it usable for takeoff. Because of noise abatement procedures, 2,500 feet must be flown over before touching down. This precious waste of concrete is found at many of the metropolitan terminals.

Runway length is not the only fault. You can drive for hundreds of miles on a turnpike and scarcely feel a bump, but landing on one of JFK's runways resembles a ride on a roller-coaster. While most airports have dangerously few runways, the Port Authority at Kennedy years ago decided to close three of its 6,000-foot runways simply because they no longer fitted into the airport complex. Parts of two of these strips were used for costly taxiways, the other as a freight-loading ramp and parking lot. The construction of these runways cost the taxpayer millions. This fact cannot be emphasized too strongly.

For over thirty years ALPA has been hammering away at Congress and the FAA to license airports and their managers. The Federal Register Vol. 34 Advisory Circular Checklist and Status of Federal Aviation Regulations lists 152 advisory circulars pertaining to airports.

These circulars, as the name implies, are advisory, not mandatory, in nature, and *they are the only Federal documents that cover airport "Standards."* There are no requirements in the FAR (Federal Air Regulations) to maintain the airport in conformance with any standard. Since all flights originate and terminate at airports, their adequacy is of prime importance.

In 1971 the FAA finally broke down and decided to license airports. Their regulations are far too lax, however. They make no mention of the minimum length of runways nor do they have any criteria for mandatory glide slopes or control towers. Marginal airfields will remain so, and your chances of a safe trip will be questionable.

The past two years have produced a slight improvement in landing aids. The FAA's Fact Book lists the following number and types of aids that have been installed:

5 Airport surveillance radars (ASR)
10 Visual oral ranges (VOR)
90 Nondirectional beacons
42 Towers

36 Airport surveillance radars (ASR)
90 Instrument landing systems (ILS)

It might seem that the FAA has really been busy making all the air carrier terminals safer, but in reality they have not scratched the surface. Eighty percent of all the runway ends at airline airports are still without the benefit of ILS. And as far as the Feds are concerned there will be very few more installed.

Perhaps by 1978 the number of runway ends served by an ILS will rise to 30 percent of the total, but there it will stop. The FAA still uses the criterion that each airport must have a given number of aircraft movements per year in order to qualify for an ILS. This ludicrous criterion leaves the safety of the airlines on very shaky ground. One-third of the airline airports still don't have the luxury of a control tower.

In the past three months I have been the victim of a very unsafe condition. More and more airports are installing the life-saving ILS, but not without a penalty. Some metropolitan airports have ILS at both ends of the same runway, so the FAA has used the same frequency for both directions. There is nothing wrong with this installation except for the real possibility of an occasional human error. When the controller switches on the localizer for one of these runways, he must also turn on the proper glide slope. Herein lies the fault.

I was making an approach at Newark Airport and the weather was anything but great. By the time I reached the outer marker the glide slope was showing a full scale deflection down. It should have been reading on course. I checked the co-pilot's gauge and he was showing the same. We received no flags or warnings for the glide slope receiver was being powered, but unfortunately not from the correct direction. We questioned the tower operator to see if he had any glide slope warning. We immediately got a series of flags and alarms as the correct glide slope sprang to life. The controller had forgotten to switch on the appropriate beam. We had no problem for we had maintained our intercept altitude; all we had to do was make a kamikaze descent and catch the glide slope. The same incident occurred two weeks later.

For more information about landing aids, please refer to page 37.

Chapter Eight

TWO STRIKES AND YOU'RE IN

IT WAS on April 4, 1927, that Colonial Air Transport inaugurated the first airline service in the United States. The initial run was between Boston and New York. In clear weather there was no problem, but when the ceiling began to drop, operations terminated. Only primitive instruments were used and flying blind was not yet possible. Pilots flew their aircrafts from point A to point B by looking out the window and checking prominent features on the ground.

Driving a car, sailing a boat, or maneuvering any earthbound vehicle requires skill in going forward and turning left or right. Flying adds a third element: up and down. When a pilot can use the horizon as a reference, staying right side up is relatively simple. But when clouds or fog erase the visual reference, there isn't a pilot alive who can maintain straight-and-level flight. Without the aid of instruments a pilot most assuredly would spin to his death.

In 1928, instruments were developed and gave the pilot a horizon to follow. A gyro was constructed that enabled the flyer to keep a miniature horizon in front of him at all times. This was appropriately called an "artificial horizon." It was, and is, the most important attitude instrument in the cockpit.

On September 24, 1928, James H. Doolittle made the first successful instrument landing. He had the same primary attitude instruments used today: an altimeter, directional gyro, airspeed indicator, and an artificial horizon. He received directional guidance from a radio course aligned with the airport runway and distance from the field was signaled to him by means of radio markers.

The tremendous technological development of the airplane far outdistances advances made in the instruments and radio aids necessary to make all-weather landings feasible. Many of the instruments in today's jets are only slightly better than the ones used by Doolittle in his open-cockpit biplane. (More on this in Chapter Fifteen.)

In 1930, a low-powered low-frequency radio was developed that enabled the pilot to home-in on the radio station. He would tune his radio receiver to the proper station ahead of him and a needle on his instrument panel would point to that station. If the needle was offset to the left of the airplane's nose, a turn in that direction would swing the needle back to the fore-and-aft axis of the aircraft. The same would apply if the pointer swung the needle back to the right side of the nose. This low-powered radio beacon was nondirectional. It would point only to the station; it gave no guidance in the form of an aural beam, and no information as to height. The beacon was of low power in the same frequency as home AM radios and highly susceptible to static. During electrical disturbances such as thunderstorms, the pointer might swing rapidly back and forth between the direction of the storm and the radio station. Another discrepancy, known as "shoreline effect," occurred because radio waves change in direction when they pass from land to water. Thus the beacon could be "off" as much as 30 degrees when tuned in by a pilot who was flying over water. Hilly terrain was another geographic disturbance. Mountains could produce a fluctuating needle, caused by the radio waves that were bounced off them.

And, since radio waves are reflected from the ionosphere in much greater strength at night, flying after sunset caused this all-important needle to give erroneous readings. Weather also had its telling effect on the unshielded antenna. Under the influence of wet snow, for example, the wire got coated and the already short range of the beacon could be cut in half.

By now you are probably wondering what this ancient nondirectional radio beacon has to do with jet-age landings. I would like to say very little, but that would be a lie. Some airports use this vintage beacon as their only approach. Under certain conditions the lives of 350 people in a 747 could depend on this highly unreliable radio landing aid. Not a comforting thought, but nonetheless true.

In November 1969 a Lear jet owned by the Mack Truck Corporation crashed while making a beacon approach into Racine, Wisconsin. With little more than a smattering of directional guidance and no glide slope at all, it is a wonder that the FAA sanctioned such an approach.

On May 2, 1970, an Overseas National Airways DC-9 ditched in the ocean thirty miles east-northeast of St. Croix, Virgin Islands. Twenty-three people died in this uncalled-for tragedy. The flight from JFK to Juliana Airport at St. Martin was uneventful until reaching the destination airport. The weather was worse than reported to the pilot, with heavy rain blotting out most of the ground. The captain attempted three beacon approaches. Each time he failed to get into a proper landing profile. Precious kerosene was used during the approaches and now the fuel remaining to reach his alternate field was marginal. The usual headlines were given to this accident: *Pilot Runs out of Fuel. Ditches in Ocean.* The National Transportation Safety Board also made their standard "Pilot Error" report. In the analysis and conclusion they stated "If the flight had proceeded to San Juan the accident obviously would not have happened." I would like to add my own statement: If the airport was equipped with a precision approach rather than a Mickey Mouse beacon, the pilot would have landed on the first attempt.

There are other ways to get a plane safely through the weather and onto the runway. Starting from the most horrendous and progressing to the safest they consist of the following approaches:

NDB	Nondirectional beacon
CIRCLING	When a straight-in approach isn't possible
VOR	Visual oral range
ASR	Airport surveillance radar
VOR-DME	Visual oral range with distance measuring equipment
ILS-BC	Instrument landing system-back course
PAR	Precision approach radar
ILS	Instrument landing system
ILS-PAR	Instrument landing system with a precision approach radar monitored

All airline airports have what is known as "approach plates." These diagrams contain a wealth of information, the specified flight patterns that must be followed to the letter if disaster is to be averted. The approach altitude is depicted, the proper course headings, the time in minutes (or seconds) from the approach facility until you can expect to see the runway. Explicit altitudes must be maintained at each phase of the approach. Descent limits are based on the type of approach utilized. The highest limits are associated with a beacon approach, the lowest is found on the best approach, the ILS. The usual limits for an instrument landing system are 200 feet and one-half-mile visibility. This varies, depending on the components in the landing system that are inoperative, the type of approach lights, and the terrain. Explicit and mandatory missed approach procedures are depicted on each plate, and must be followed exactly.

The whole procedure on each plate is there for one reason only: to make a safe landing. A pilot goes through all the turns and descents that the diagrams show and ends up on the final approach at the prescribed minimum altitude. Here he stays and drives onto the airport. If he sees the field, and if he is in a good position to land, he does so. If the runway is not in sight when it should be, an immediate pull-up is made and perhaps the pilot will try again. If a second try fails, the pilot flies to an alternate field where the weather is more suitable for landing.

On November 10, 1971, the NTSB, investigating an incident at Gulfport, Mississippi, disclosed a startling FAA error. While tracking down the clues that made a Southern Airways DC-9 hit some power lines, they discovered that an approach procedure for an ILS to runway 13 had been published seven months before the accident. There is not, and never has been, a commissioned ILS at that airport.

There are two broad approach categories. The most dangerous are called "nonprecision," that is, they lack the vertical guidance of a glide path, and only tell direction. Some examples of this type are VOR, Localizer, NDB, Back Course, Circling, and ASR. The only instrument between you and disaster is a pressure altimeter; this instrument has given erroneous information before, and will again.

The Circling approach is on a par with the beacon let down. The alignment of the runway with the final approach course determines if a straight-in can be executed or not. Anytime the extended centerline of the runway and the navigational approach aid differ by more than 30 degrees, a circle to land must be made. The first portion of this approach is done as if a straight-in landing were to be made. Many times, because of the wind direction, a straight-in is not feasible. The pilot follows the normal procedures to the field and descends to the published approach minimums. Now the field must be in sight and a turn to a downwind leg is executed. On a rainy night, it is bad enough to make a straight-in nonprecision approach but to circle is dangerous.

The letdown plate may be geared for a circling approach, but believe me, the airport isn't. Runway lights are directional. The greatest amount of light is fed up and down the length of the runway. When a runway light is viewed at 90 degrees to the runway, it is barely visible. Yet this what the pilot has to gauge his landing on as the ground visual cues are all he has. This is primitive flying at its worst, and should not be allowed in a prop, let alone a jet. The terrain clearance for this approach is far too tight for comfort. Add altimeter error plus lag and an altimeter setting that may be received from a radio station miles away, include the mountainous terrain, and you have the ingredients for disaster. American Airlines does not allow their pilots to make circling approaches unless the weather is VFR. This doubles the safety factor that the FAA says is legal. All airlines should follow suit.

On May 9, 1970, a Lear jet crashed near Emmet County Airport, Michigan, killing all on board. United Auto Workers' President Walter Reuther had his life snuffed out by a circling approach and a faulty altimeter. The scene was the usual rainswept airport with an approach that lacked any visual clues at night except for a few lights from widely scattered houses. The area was thickly wooded and there were no approach lights on the runway. The following are a few other crashes that have taken place in the last few years. All of the captains were making nonprecision approaches with no glide-slope guidance.

November 14, 1970:

a Southern Airways DC-9 was shooting a localizer-only approach into Huntington, West Virginia. All seventy-five on board perished. When Mr. Shaffer, chief of the FAA, was questioned by a number of senators at a Senate Appropriations Committee meeting concerning the difference a precision approach would make, Mr. Shaffer's reply was: "An additional visual clue would have constituted no guarantee that the pilot would not have dropped below the slide path." With this type of reasoning, it is a wonder that there are any glide slopes at all.

September 15, 1970:

an Alitalia Airlines DC-8 broke in half after a hard landing at JFK. The jet was turned too close in to the runway. The pilot should have been at least seven miles out, but he was vectored as close as three miles. He was too high when he reached the runway. Since there was no glide slope to aid him in planning his descent, he tried to save the approach by using reverse thrust in the air. The plane broke in two when she hit. Luckily no one was killed.

June 7, 1971:
an Allegheny Airlines Convair 580 was shooting a VOR approach in foul weather at New Haven's Tweed Airport. The craft hit less than a mile short of the runway; twenty-eight people perished in the fiery wreckage. There were no approach lights nor a glide path at this airport.

June 22, 1971:
a Northeast Airlines DC-9 struck the water while executing a VOR approach to Martha's Vineyard, Massachusetts. Due to the skill of the captain, only the back part of the aircraft sustained any damage as it touched the water. An immediate pull-up was made and a successful landing followed at Boston.

September 4, 1971:
an Alaska Airlines Boeing 727 slammed into a rocky cliff eighteen miles short of the runway. A nonprecision-localizer-only approach was being used. The 111 people who died will never get to use the temporary-distance-measuring equipment that the FAA recently installed.

The Juneau tragedy: the worst plane disaster in U.S. commercial aviation history brings to twenty the number of fatal accidents that have occurred during nonprecision landings in the past decade. Nearly seven hundred people have died in these accidents.

Precision approaches are the finest: they contain guidance in direction and altitude either with a glide path associated with a full ILS, or with the guiding voice from the final controller on a precision radar approach. There are approximately nine different types of approaches, but only two of them are precision.

The VOR is used at many terminals as the only means of getting to the runway. The radio signal emitted from this facility is very high frequency (VHF) and is supposedly not affected by static. This ground transmitter projects 360 different courses a pilot can fly; however, for an instrument approach, only two or three are utilized. The reception range is governed by line of sight. The higher the airplane, the further away it can pick up the signal from the VOR. If a mountain gets between the transmitter and receiver, the signal will be unreliable. Weather plays havoc with VOR stations: the buildings that house them have flat roofs so as not to interfere with transmitter signals. Any irregularity in the signals emitted will trigger an alarm which shuts down the VOR. So sensitive is the alarm that a good snowfall will deflect the beam and turn off the system, usually when it is most needed.

Another drawback is the limited number of radio frequencies allotted for VORs. It is not unusual to select a VOR station in the cockpit and have the DME (distance measuring equipment) feature of it channeled to a nearby station, giving a completely erroneous mileage from the station. The aircraft receiver is no different from most: it dislikes hard work and will lock onto the nearest signal.

When all components of the VOR are functioning properly, a pilot can set any course desired and fly it. Once the needle in the airplane's receiver is centered, a pilot can stay on the course selected by keeping the needle centered. If the needle wanders to the left, the pilot turns left in order to recapture his proper course. Another function performed by this instrument is a positive indication when passing over the VOR station. A small pointer covers the space marked "to" when going toward the station, and flicks "from" when passing over it. At times the RPM setting on some propeller-type aircraft can cause the needle to fluctuate. A change in the engine setting will usually smooth out this roughness.

Nevertheless, this navigation aid still gives the pilot direction only; there is no way of knowing if he is maintaining a correct descent angle. An added feature accompanies the VOR at some ground sites. It is the DME (distance measuring equipment). On aircraft equipped to receive DME, the pilot can read directly from this instrument the exact mileage from the station. This feature should be made mandatory for all VOR stations.

On December 24, 1968, Captain Gary Lee Mull, thirty-three, was in command of Allegheny Airlines turboprop Convair Flight 736. Mull had been promoted to captain in 1967, after serving as a first officer for three years. He was known as a highly capable flier, and a safety-conscious one.

Flight 736 originated in Detroit, destined for Washington, D.C. One of its scheduled stops was already completed, at Erie, Pennsylvania; the next landing was to be Bradford, Pennsylvania. The Bradford airport has many scheduled airline flights daily, yet it still has a second-class landing aid, the VOR.

This Christmas Eve was just the way most people wanted it, white. Flight 736 had a happy group on board. The Christmas spirit was everywhere, and stewardess Rita Boylan remarked to Captain Mull, "I sure wish all our passengers were in such good spirits all the time."

Captain Mull said, "Well, only three more ups and downs and we'll be home with our families. Just got the weather for Bradford. It's typical blowing snow, low ceiling with one mile visibility. We'll be there in fifteen minutes, so better get them ready."

"OK, Gary," she replied and left the cockpit to attend to her duties.

First Officer Richard Gardner was busy working the radio: "Hello Erie 736, do you have any specials for Bradford?"

Erie replied, "736, the Bradford weather estimated 2,000 broken, one mile visibility, light blowing snow, wind 310 at 15 to 25, altimeter 29.77."

(Any weather that is "estimated" can mean it was taken by a guy who looked up at the sky and guessed the height of the clouds. Highly inaccurate.)

Erie again called 736: "What is your position right now?"

Gordon replied, "About 4.5 miles from Bradford VOR."

(Erie, the controlling facility for Bradford, is in charge of all IFR operations at the field. They have no radar, yet the FAA sees fit to allow them to space traffic.)

Erie said, "You want to go over to Bradford radio now?"

Dick said, "OK, will do, will see you then. Bradford radio, Allegheny 736."

"Roger, Allegheny 736, understand you're over the VOR starting approach. Runways 14 and 32 are covered with hard-packed snow and rough ice braking poor by a Convair, sir."

"OK, 736."

The final cockpit check was now completed. The stewardess checked the seat belts of her forty-four passengers. Flight 736 was outbound from the beacon and in solid-instrument conditions. (In other words, they were flying completely on instruments, no ground contact. The NDB that the captain said he was past was used in conjunction with the VOR. This is the same type of beacon discussed at the beginning of this chapter.)

Captain Mull was in the process of giving a final look at his approach plate when he said to his co-pilot, "Better give me a little flaps there—just about fifteen."

"Fifteen of them coming at you."

"OK, Dick, drop the gear."

"Yeah, comen-atcha: seat belt, no smoking, engine deicing. You're about 2.5 miles from the airport."

Still in a solid-cloud condition, Flight 736 was at minimum altitude 500 feet above the field elevation. (Now the time for decision is at hand; if the lights of the runway are not visible soon, a pull-up will have to be initiated.)

Captain Mull flipped on the landing lights and said, "If it will help to see it, I don't know."

Dick replied, "I don't see a thing."

"PULL UP!"

At 8:11 P.M. Allegheny Flight 736 plowed into a forest, snapping trees like toothpicks. Still losing momentum, it traveled 1,000 feet farther before ground contact. Finally, its energy spent, it flipped over on its back and came to rest. Both pilots were killed instantly, along with eighteen others.

The report of the NTSB stated:

> The probable cause of this accident was the continuation of the descent from the final approach fix through the minimum descent altitude and into obstructing terrain. Contributing factors were the minimal visual references available at night . . . a rapid change in visibility conditions that was not known to the crew.

On January 6, 1969, Allegheny Flight 737, turboprop Convair 580 (similar to the one that had crashed thirteen days prior) slammed into a forest while making a VOR approach at the same airport. Nearly all the conditions of the first crash prevailed. This aircraft was also groping for the runway on a snowy evening, and ended up inverted, killing the crew and nine passengers. Even the time of the accident was nearly identical to that of the first disaster, taking place only twenty-three minutes later. The NTSB has not decided on a probable cause for this second accident. They did make some recommendations to the FAA.

Here are some of the pertinent statements the NTSB made in their safety-information pamphlets. My comments are in italics.

> The chief of the Aerospace and Procedures Branch, Flight Standards, FAA, made the following testimony. He described the minimum requirements for the installation of an Instrument Landing Systems (ILS). He further stated that Bradford meets these requirements and has for some time. Bradford is programmed for the installation of an ILS, along with an Approach Lighting System (ALS), in the near future.
>
> *Eleven months after the first accident, Bradford got their much-needed ILS.*
>
> A letter from FAA to the NTSB stated that Bradford met the criteria necessary to qualify for the installation years before the accident. However, budgetary restrictions have limited the rate at which ILS can be installed even at those airports which qualify.

In 1968, Congressman Fletcher Thompson of Georgia assailed the FAA extravagant and unnecessary expenditures. He charged that the FAA spent over $3 million in the past three years to rent airplanes over and above their permanent fleet of 100 planes. Personal flights are on record.

Three million dollars buys a lot of navigational hardware. The average cost of an ILS system is approximately $150,000.

The Assistant Director of the Collins Radio Company discussed the design of the radio receiver that they manufactured and installed in Allegheny aircraft. He discussed the chain of circumstances necessary to cause an erroneous indication on the airborne equipment. With only a few such occurrences reported he described this as unusual but possible.

In recent months there have been numerous radio receiver discrepancies. I have had some myself, and there are many on record.

Another contributing factor to the accident was the "Minimal Visual References" available to pilots for night approaches to Bradford.

There are still no approach lights installed at Bradford. These high-intensity lights extend 3,000 feet out from the end of the runway. They afford an excellent means of lining up, especially during inclement weather. There are other cheaper yet effective lights that could have been installed such as runway end identification lights (REIL) but no improved lighting has been added.

Another contributing factor "a rapid change in visibility conditions that was not known to the crew." *In fact,* the weather conditions in the final approach area may have been as low as 800 feet variable and one half mile visibility. This report was taken 2 minutes after the crash.

Vital weather reports are still being taken by the people that load baggage onto the plane. They are given a few days' training and after a simple exam are licensed by the Government to make weather observations. Meteorologists go to college for years and they make mistakes.

That a ramp agent could handle such an important job is hardly likely. Estimated ceilings should be abolished. At times they are still taken this way at major airports such as Kennedy, O'Hare, and Los Angeles, and they are always in error. The weather observation taken two minutes after the crash was below the legal limits for the airport by one-quarter of a mile.

The report also stated the co-pilot failed to call out the altitude at 500 feet above field elevation, as required by company regulations. The board noted, however, that it was possible he was looking outside at the time to meet another Allegheny requirement to call out ground reference.

It is impossible to devote full attention to calling out altitudes which requires a constant vigil inside the cockpit and at the same time expect to see the ground without looking for it.

When Bradford airport received their ILS, the minimum visibility allowable was one mile. A few weeks later, with no additional landing aids, the limits were lowered to three-quarters of a mile. There is still no control tower at the field and no radar.

On October 25, 1968, a Northeast Airlines Fairchild 227 turboprop crashed at Hanover, New Hampshire, killing thirty-two of the forty-two on board. Another VOR approach was involved, but this time the added safety feature of DME was missing. The VOR at Lebanon is a straight VOR which puts it nearly in a class with the primitive beacon approach. The Lebanon weather was an *estimated* 2,000 feet, overcast, with 10 miles visibility.

One of the surviving passengers had this to say about the weather: "We were so near the ground at this time that I could clearly see the individual trees and as they seemed to get near enough to touch, their view was blocked out by clouds."

Two forecasts from different agencies both predicted ceilings between 200 and 400 feet with a visibility of 2 to 3 miles in fog. This forecast gave a truer picture of the actual weather.

In 1961, the minimum safe altitude within 10 miles of the VOR was 4,500 feet. The altitude of the VOR was 3,300 feet. In January of 1966 these were both lowered. The 10-mile limit was reduced from 4,500 to 4,200 feet and the VOR altitude was lowered to 2,800 feet, 500 feet less than previously. No upgrading of the airport aids was noted. Apparently a "small" mistake made by the FAA went unnoticed for five years. The old time limits exceeded the maximum allowable descent for an approach and caused the pilot to be too high for a landing.

The NTSB fixed the probable cause of this accident as the premature initiation of a descent toward minimums based on navigational-instruments indications of an impending station passage in an area of course roughness.

Since the accident, the state of New Hampshire has installed a low-frequency beacon. Yes, the same type of beacon first put in use forty years ago. The FAA approved this innovation, and because of this vintage beacon part of the procedures were lowered 300 feet. This in effect took away any of the newly arrived safety benefits.

The investigation disclosed that the Lebanon VOR was unreliable and gave incorrect indications to flight crews. Eight incidents of partial or complete station passage were indicated prior to actually reaching the station. Some as far out as ten miles. The latest reported malfunctions were on June 10 and October 27 of 1968.

Joseph J. O'Connell Jr., chairman of the investigating committee, sent a letter to the acting administrator of the FAA, D. D. Thomas. Pertinent parts of that letter said: ". . . Two of Northeast's aircraft reported station passage prior to actually passing the VOR. This happened both before and after the accident. TWA reported a number of erratic operations. Two false reversals were actually observed by two of your FAA inspectors. . . ." Both carriers were equipped with Wilcox VOR model 806A aircraft receivers.

Mr. Thomas's answer to this serious problem consisted of a directive that all users of this navigation receiver be aware of its idiosyncrasies and use caution.

Tests were performed which proved conclusively that the Wilcox receiver consistently gave more erroneous indications than any other receiver. They still were not ordered repaired by the FAA. Instead, an advisory circular was distributed in which a warning was posted concerning certain radio equipment; but the warning did not specifically mention the tested and proved culprit. On January 1, 1969, the Wilcox Electric Company issued a service bulletin which *recommended* three modifications to improve the performance of their receiver.

Another letter, dated December 13, 1969, was sent to the acting FAA administrator from J. J. O'Connell, chairman of the NTSB. It contained some interesting statements.

> It was disclosed that *prior* to the accident a Northeast flight crew had experienced a false indication of station passage while making an approach to the Lebanon Airport. In this incident, the crew was completing the procedure turn inbound when the course needle fluctuated and the To-From indicators went from "To" to "From" indicating station passage.
>
> With these indications, the crew started a descent from 2,800 feet. Upon reaching 2,000 feet the crew then noticed that the To-From indicators had reversed, indicating a 'To.' The Captain observed the nearness of the terrain through the breaks in the overcast and immediately applied power and climbed back to a safe altitude. . . .
>
> This incident was reported to the local FAA maintenance technician who initiated a routine check of the facility which uncovered no irregularity. However, he did not, nor was he required to by your current procedures as we understand them, report this occurrence to any central unit within your organization.

On January 14, 1969, Mr. Thomas replied that federal regulation 121.561 requiring a pilot to notify an appropriate ground station of any navigation irregularity was a good one and that no change would be made.

If this regulation is to be a part of the FAA's lip service it most certainly is useless. A tragedy could possibly have been averted—if such highly important reports were not allowed to terminate with local maintenance men. The pilot who submitted that discrepancy was complaining about a deadly serious condition that existed in the air. No amount of tinkering with the tubes on the ground can ever duplicate a bad reading in the air. That facility should have been taken off the air until a thorough flight check was performed.

Mr. Thomas went on to say: "The Lebanon VOR performs within those tolerances and, therefore, should not require an additional facility to support the instrument procedure. Notwithstanding budgetary constraints, we would like to see a DME located at every VOR site. However, our ultimate objective is to provide vertical guidance as well as directional, at all air carrier airports."

The FAA continues to waddle in a pool of inefficiency. If idle talk were gold, the FAA would be prosperous enough to give money away.

The FAA's bird study in the mid-1960s is a classic example for collectors of government trivia. After months of research and thousands of dollars spent, the FAA was convinced that the building of an airport near Huntsville, Alabama, would not be safe. The proposed site was only a mile from a 38,000-acre wildlife refuge; over 100,000 geese called this their home. In April 1966 Congressman Brooks of Texas questioned Mr. Thomas about why, after spending so much money to prove the airport site dangerous, the FAA spent $4.5 million to build an airport there? Mr. Thomas admitted that the airport was constructed on the advice of some local duck hunters.

As for the statement that the FAA wants a precision approach at Lebanon, New Hampshire, absolutely nothing has been done since the Northeast tragedy. And that was nearly four years ago. A beacon was placed on the approach course by the State of New Hampshire. All the FAA did was to lower the minimums and thereby decrease safety.

On November 19, 1969, a Mohawk airliner Fairchild 227 turboprop crashed into a mountain while executing a VOR approach at Glens Falls, New York. All fourteen persons aboard were killed. No improvements have been made at this site. Both Glens Falls and Lebanon are airports that have minimal facilities. No tower, no radar, no DME, no glide slope, no approach lights.

Nearly every aircraft system has a series of backups. If one fails, another can be used. When a navigation aid is inoperative and it is the sole means of assisting aircraft to find the field, the airport is automatically closed. It is extremely important that all navigation facilities be monitored. The controlling facility for en route aids receive a warning signal whenever a malfunction occurs. The ILS has monitors in the control tower that warn of a fault in any component. However, numerous facilities are unmonitored; the only way they are detected is by a pilot's report—hopefully before it is too late. Airline airports should be equipped with a minimum of two separate approaches, preferably ILS approaches.

In my opinion, all of the crashes mentioned in this chapter could have been avoided if the airports were ILS and radar equipped. Pilots plan for the unexpected, and having an ace in the hole is part of safe flying. On a VOR approach you have no backup. Three airlines have curtailed the use of a VOR-DME approach, and have issued restrictive orders to their flight crews. The government has done nothing. The current FAA criteria for installing the ILS are shocking. It would seem that two airliners must come to grief at the same field before minimal equipment can be installed.

That's the way Bradford got theirs.

Chapter Nine

THIRTEEN SECONDS

WHEN a pilot on an ILS breaks through the clouds and scans the outside world for sufficient visual clues to execute a safe landing, he has exactly thirteen seconds to make a life-or-death decision. (At a few airports this thirteen seconds has been cut to six because of lower landing limits.) The captain has thirteen seconds to decide if 282 tons of metal traveling at 140 miles per hour can be gently and safely planted on the runway. Add to this the fact that information portrayed by the instruments in a modern jet becomes less accurate as the aircraft descends from the 200-foot point, the minimal allowable descent altitude unless a definite ground reference is seen.

When executing an instrument approach, the captain usually flies the aircraft solely by reference to his instruments. The first officer and flight engineer run the check list. When the first officer is not busy with duties inside the cockpit, it is his job to watch for visual clues. The state of the art has not perfected blind landings, so it is still necessary to land the plane without reference to the instruments. As soon as the F/O spots the approach lights, he calls, "I have the lights."

At this time the aircraft commander raises his eyes from the instrument panel and prepares to make a normal visual landing. The eye is not capable of too-rapid focus; it has been proved that the human eye takes 2.39 seconds to shift from short range to long range. A pilot who has for hours been scanning an instrument panel two feet in front of him has difficulty changing his range of vision to scan great distances. The actual flare and landing are usually accomplished best by senior pilots —not because they are endowed with special skills, but they have been at it longer.

Now that we have discussed the landing procedure, let's examine the components of an instrument landing system: localizer, glide slope, and outer and middle marker. These four components are minimal for a full ILS system.

Localizer: a radio transmitter placed at the far end of the runway to give the pilot directional guidance between 3 and 6 degrees.

Glide path: a radio transmitter approximately 1,000 feet down and to the side of the runway. It guides the plane down to the strip at an angle of 3 degrees, depending on the local terrain.

Outer marker: alerts the pilot as to his distance from the field. It is usually placed between 4 and 7 miles from the end of the runway. As the craft passes over the outer marker, a purple light flashes in the cockpit; an oral signal is received as well.

Middle marker: identical to the outer marker except for its location. The middle marker is approximately 3,500 feet from the end of the runway. When a pilot overheads this marker, an amber light is seen, signifying that minimum descent altitude has been reached.

When all components of the ILS are functioning, the minimum visibility allowable for landing is three-quarters of a mile. When approach lights are installed and operating, the visibility drops to one-half mile. The more aids you have, the less the visibility requirements. If any of the various components become inoperative, the visibility requirements are raised. For instance, if the glide slope of an ILS system becomes inoperative, the minimum allowable visibility for landing increases one-quarter mile. In essence an ILS that has lost its glide path is no better than a VOR approach. When certain wind conditions prevail, approaches are made using the oldest and most unreliable of all navigation aids, the radio beacon. Most ILS installations include one of these low-frequency radio beacons in conjunction with the outer marker. It adds another backup to the system, and backups mean safety.

An ILS is no different from any other system when it comes to reliability. Heavy wet snow, or even a soaking rainstorm, can put the localizer and glide path off the air. The glide path is the more critical of these two components. Rather than reflect its electronic ladder directly, it is bounced off the ground a few feet in front of its transmitter. But the earth is subjected to the elements, and this changes the electrical ground for the reflected glide path.

A pilot making an ILS approach has only one backup system. He can check his indicated altitude over the outer marker as compared with the altitude printed on his approach plate. The altimeter, glide slope, and marker should all agree that he indeed is where he thinks he is. When the aircraft passes over the outer marker and begins its descent to the field, the only backup system for the glide slope is the altimeter.

Precision approach radar (PAR) is so accurate that if an aircraft descends as little as five feet below the prescribed glide path, it is noticed on radar and a corresponding warning is given by the controller to the pilot. This system of letdown was widely used in World War II, and is sometimes called "ground control approach" (GCA). When this system is used in conjunction with an ILS, it makes the safest system yet devised.

On November 20, 1967, a TWA Convair 880, under the able command of forty-five-year-old Captain Charles L. Cochran, was making an ILS approach to the Greater Cincinnati

Airport. At 8:57 on this cold and stormy night, Flight 128 terminated 7,000 feet short of the field. Only thirteen of the eighty-two people on board lived.

Because of the lengthening of the main runway a number of the ILS components were shut down. (It isn't uncommon for vital components of the ILS, or even the whole system, to be closed down for months, and in some cases, years.) The approach lights, the middle marker, and the glide slope were all off the air. Available were the localizer, outer marker, radio beacon, and surveillance radar, a type of radar that has no height-finding capabilities and is directional only.

Cincinnati Airport over the past few years has been host to a number of landing incidents. All at night, all in bad weather, all making nonprecision approaches.

The NTSB final decision concerning the crash: "An attempt by the crew to conduct a night visual, no glide slope approach during deteriorating weather conditions without adequate altimeter cross-reference." The cause depicted by the board tends to put the blame on the crew. They were following their FAA approach to the letter. As long as such dangerous letdown procedures are sanctioned by the government, disasters are inevitable. A brief resumé of some air carrier accidents follows:

Piedmont Airlines, F-27, Charleston, West Virginia, August 10, 1968: accident occurred during an approach to landing; plane crashed 250 feet short of the runway.

Japan Air Lines, DC-8, San Francisco, California, November 22, 1968: plane struck the water and landed in San Francisco Bay.

Trans Texas Airways, DC-9, Austin, Texas, December 26, 1968: plane clipped an approach light 400 feet short of the runway.

Scandinavian Air Lines, DC-8, Los Angeles, California, January 13, 1969: plane struck the water and landed 6 miles west of the field.

None of these accidents should have happened. None of the crews reported any drastic malfunctions. They were all victims of FAA ignorance. In all of these accidents a precision approach radar controller could have saved many lives by warning the crew that they were getting dangerously low. But this necessary backup aid was not available.

The ILS is by no means a new marvel of the electronic world. It was first used over twenty years ago and only minor improvements have gone into the system. Passengers in a 747 are still sliding down the same ILS system that was used by the DC-3. Today's ILS is not even as safe as it was then.

In October 1965 the FAA decommissioned all ILS compass locaters (an aid that gives a double check when overheading the marker). This was fought by ALPA, but the FAA was saving money. In 1963 there were thirty precision approach radars in operation; not enough, but better than the latest count. Among airports without PAR: Pittsburgh, Pennsylvania; Washington, D.C.; San Francisco, California; Boston, Massachusetts; and JFK, New York. During the latter part of 1970 the FAA tightened the purse strings again and decommissioned all existing precision approach radars. Now when your pilot is on the final approach, he is at the mercy of his inaccurate instruments. No radar monitoring is available. Only 10 percent of the 3,000 runway "ends" used by airlines have full instrument landing systems.

Canada is a country that cannot compete with the United States when it comes to air traffic, but it has better equipped airports. Los Angeles International Airport is the only field that can compare with Toronto. Both of these airports have some form of ILS on every runway. Toronto has precision approach radar to monitor landings for each of its four runways—and Toronto works exactly half the traffic that JFK does. It has been fifteen years since the ILS was called necessary equipment at airline airports. Today, approximately 300 of the 530 airports served by U.S. air carriers still do not have ILS.

Just how *does* an airport get a landing aid from the FAA? Well, the Federal Aviation Administration does a lot of *talking* about safety. The safety criteria for obtaining approach aids is based on the amount of traffic at each individual airport.

I'm not suggesting that every grass strip across the nation install a precision approach. But when the lives of millions of passengers are jeopardized at ill-equipped airports, it is time for the FAA to get a new standard.

Chapter Ten

SEE OR BE DESTROYED

AIR TRAFFIC regulations were first written over thirty years ago; the majority of those rules are still in effect—unchanged. In the days of the DC-3 the FAA's "see and be seen" rule was adequate. The closing speed of two DC-3s was 300 miles per hour. Today's jets close at speed four times greater.

The criteria for VFR flight was, and still is, 1,000-foot ceiling (lowest cloud deck) and 3-mile visibility. There are two basic types of flight plans: IFR (instrument flight rules) and VFR (visual flight rules). IFR is generally used when the weather is bad and ground contact cannot be made; VFR is used in good weather. All commercial turbojet aircraft must file IFR flight plans regardless of weather. In spite of the fact that IFR gives you the protection of controlled separation, it is still the responsibility of the flight crews to see and avoid other aircraft in conditions that will allow it.

Aircraft are normally spaced a minimum of 1,000 feet vertically when flying assigned routes in their private block of air. "Private" is really not the right word, because also contained in this assigned safe route are hundreds of other craft, mostly general aviation, that can travel to and fro, up and down, as they see fit. All within the law and under no one's control. Ground level to 3,000 feet is the pilot's "no man's land." Here is where the majority of the little fellows practice their stalls, turns, and various other maneuvers, completely oblivious to any other traffic. VFR pilots can fly on any heading of the compass up to but not including 3,000 feet.

It wasn't too long ago when pilots felt safer in their planes than in their cars. That's all changed now. During the period from 1956 to 1967 a survey of pilot deaths showed: 39, auto accidents; 69, cancer; 143, coronaries; 197, air carrier accidents. A quote from an ALPA safety chairman speaks for itself: "As an airline pilot, your best chances for dying are no longer from injury suffered in an auto mishap while attempting to make your way to the airport. Cancer or heart attack isn't number one either. *Mid-air* gentlemen, that's what it's gonna be. . . ."

Chapter Eleven

ALUMINUM SHOWERS

THE NATIONAL Transportation Safety Board (NTSB) recently released their study of "Midair Collisions in Civil Aviation 1968" which reports 35 collision accidents involving 76 aircraft, a 46 percent rise over 1967. The major problem in the 1968 mid-air collision accidents was the failure of pilots to adhere to the "see and be seen" concept. The majority of the collisions occurred at or near uncontrolled airports and below 5,000 feet. In 1938 the little fellows managed to rack up a total of 1,861 accidents and 274 fatalities. In 1967 this figure swelled to 6,115 accidents with 1,228 deaths.

The FAA has never increased the VFR criteria of 1,000-foot ceiling and 3-mile visibility. Instead they have lowered it, making the skies more dangerous by allowing flights under special VFR. This enables a pilot with no instrument experience to take off from a crowded airport and mill around, trying to find a life-saving hole in the clouds or haze and press on through. As long as he has one-mile visibility and remains clear of the clouds, he's legal. Professional Air Traffic Controllers Organization's Mike Rock says: "Special VFR is a real killer. More than once I have seen pilots take off and enter the haze only to come spinning out of the overcast. It's literally impossible for a man to utilize this procedure and call when he's clear of the control zone (usually five miles from the airport). The book says as long as he's clear of the departure zone, he's legal. No mention is made of the havoc he can cause when he leaves one zone and charges, unannounced, into another."

In 1968 there were 2,230 near misses reported across the country. (The FAA's definition of a "near miss" is when two planes are closer than 500 feet from each other and have to take evasive action to keep from colliding.) This was the first year that the pilot and controller would be immune from punishment, even if either was in error; prior to this time, any reported near miss would bring harassment from the FAA. Each year since 1968, the FAA has announced continuance of the immunity program for another year. It was extended until December 1971 when it was abandoned. The near-miss problem has been with us for years. The files they are collecting concerning near misses are a cover-up to enable the traffic system to deteriorate for still another year.

At any moment during the daylight hours the FAA estimates that there are between 9,000 and 10,000 craft aloft on the skyways, as many as 5,000 of them on the "Golden Triangle" formed by lines connecting Chicago, Washington, D.C., and New York. With 1,000 new aircraft a month being added to the nation's aerial hardware, the traffic will soon come to a standstill.

On June 30, 1956, a TWA Connie and a United DC-7 collided over Grand Canyon, Arizona. One hundred twenty-eight people perished in this, the first of the famous mid-air accidents. Both planes were flying at 21,000 feet in uncontrolled airspace. Both were legal; the fault lay in the traffic system and the regulations, even as it does today.

TWA Flight 2 asked ATC (air traffic control) for a 1,000-foot on-top clearance. This gave the pilot all the sky in the world to play with. At least 1,000 feet on top of the clouds could place him at a variety of altitudes. As long as both planes were not in the confines of the federal air space, no separation was expected or offered. The official CAB (Civil Aeronautics Board) probable cause was: "The Board determines that the probable cause of this mid-air collision was that the pilots did not see each other in time to avoid the collision." It was also found that: "There existed an insufficiency of en route air traffic advisory informations due to inadequacy of facilities and lack of personnel in air traffic control."

After the accident, and true to form, on December 1, 1957, the CAA instituted a continental control area above 24,000 feet. This meant that all planes would be under positive control whether they were flying on or off the airways. The regulation was mandatory under conditions of restricted visibility, but could be ignored during good weather. Well, it was better than nothing.

On April 21, 1958, United's Flight 736, a DC-7 from Los Angeles to New York, was rammed by a U.S. Air Force F-100 fighter. Forty-nine people lost their lives. The collision occurred at 21,000 feet in good visibility, at exactly the same altitude and nearly the same position as the previous accident. United was on an IFR flight plan, the fighter on a VFR training hop.

Testimony of Air Force personnel stated that Nellis AFB in Las Vegas did not perform any air traffic control function except for Nellis aircraft, and use of the function did not relieve the instructor pilot of visual separation responsibilities. Many times over the past few years, United pilots were plagued with near misses when approaching this military training area. These reports were filed with United and the CAA—all to no avail.

Probable cause? Inability of the pilots to "see and be seen," and of the CAA to take every measure to reduce a known collision exposure.

On May 20, 1958, a Capitol Airlines Viscount and a Maryland Air National Guard T-33 trainer collided in the air about four miles east of Brunswick, Maryland. Everyone concerned died, with the exception of the pilot of the T-bird, who ejected safely. Capital was on an IFR flight plan, the T-33 on a VFR training mission.

Probable cause? "The board determines the probable cause of this accident was the failure of the T-33 pilot to exercise a proper and adequate vigilance to see and avoid other traffic."

On May 28, 1958, the CAA adopted over certain routes a positive airspace between 17,000 and 35,000 feet. This rule differed from the previous regulation in that now all traffic would be on IFR flight plans, regardless of the weather. Considering the extreme closure rates of high performance aircraft, this rule would help to reduce accidents. In spite of the improved controlled air space the ancient "see and be seen" concept remained in effect. Within a year TWA would originate coast-to-coast jet service with its new 707s. Seeing traffic at the slower prop speeds was difficult enough; soon it would be next to impossible.

In 1967 extensive tests were conducted to prove conclusively that the "see and be seen" concept of flying a modern jet in this environment was unquestionably *dangerous.*

It's a well-known fact that pilots have better-than-average eyesight. It is also true that the majority of collisions occur in good weather near an airport. Why? The basic limitation is simply that a man cannot see, identify, or react to an object the instant it comes into view. This process takes time, not much time, but enough for a high-speed jet to travel thousands of feet. Speed is the limiting factor for reaction time.

A pilot flying at 600 miles per hour spots something out of the corner of his eye. He will travel 920 feet before he can definitely determine if it is a cloud, a speck of dirt on the windscreen, or another aircraft; 2,680 feet will be used up while he decides whether to climb, dive, or turn left or right; a precious 4,792 feet will slip by before he can make the plane respond to his actions. A jet pilot would need a bare minimum of 9,584 feet before he could spot a target and maneuver to miss it. And these figures were compiled under excellent conditions: clear day, eyes focused for distant vision, no distractions, and a well-rested crew.

The size and color of the target are of great importance. If the FAA demanded that specific sections of planes be painted with a bright day-glo orange, their chance of detection would be greatly increased. The cockpit of a jet airliner is a very busy place. Pilots cannot continually maintain an outside vigil at all times.

The tests also brought out the fact that the observers on the test planes, when told where and when to expect targets, had little difficulty in spotting them 12 miles away. Pilots, however, cannot distinguish traffic until it is 3 to 5 miles distant. High-performance fighters remain undetectable until 2.3 miles away. FAA rules still say that 3-mile visibility is sufficient. The safety margin is dangerously small: one-third of a mile or about two seconds.

On December 6, 1960, a mid-air collision occurred between a TWA Connie and a United DC-8 over Staten Island, New York. TWA was en route from Ohio to LaGuardia Airport and United was on a nonstop trip from Chicago's O'Hare to Idlewild (now JFK), New York. Both trips were under IFR and were being monitored by New York center radar. No difficulties were reported by either flight. TWA was cleared by center to descend to and maintain 9,000 feet; upon reaching 9,000 he was told to contact LaGuardia approach control, which he did. Now both flights were being monitored by different facilities. United Flight 826 was at 13,000 feet and had just passed Allentown, Pennsylvania, when New York center called:

CENTER: 826 cleared to 11,000 feet to proceed on Victor 30 until intercepting Victor 123 and that way to Pres-

ton. It'll be a little bit quicker. [This new routing would shorten the distance to Preston Intersection by 11 miles.]
(United acknowledged the clearance and continued their descent.)

CENTER: Looks like you'll be able to make Preston at 5,000.
UNITED: Will try.
CENTER: If holding is necessary at Preston S.W. one minute pattern right turns . . .
UNITED: Roger, we're out of 6 for 5.
CENTER: 826, Roger, and you received the holding instructions at Preston, *radar service is terminated.* Contact Idlewild approach control.
UNITED: Good day.

Simultaneously TWA was working LaGuardia approach control and was being vectored for an ILS to runway 4. At 10:33 LaGuardia approach advised TWA: "There appears to be jet traffic off your right now 3 o'clock at one mile northeast bound.

At 10:33.28 United Flight 826 called Idlewild: "Idlewild approach control, United 826 approaching Preston at 5,000."

Those were the last two radio calls made from each flight. At this precise moment the planes rammed one another, killing 128 on board, and 6 on the ground.

Mike Rock, who was working in the New York center at the time of the accident, had this to say: "I was not working radar during the time of the accident, but I know that LaGuardia radar was being worked by a trainee. We had told our supervisor time and again that an accident was bound to happen. The communication set up between the three towers and center was very poor. (New York center, LaGuardia, Idlewild, Newark.) We were using an old GRS (general railroad signal device used in railroad freight yards) to signal to another control when the various altitudes were clear. We kept telling our bosses we needed a direct interphone between the facilities. Their reply was that funds were lacking. Two days after the crash we got the drop line. It cost $75."

A new phone system was not the only result of this accident. Positive radar handoffs became mandatory. Now a controller could not relinquish his guidance of a flight until the receiving controller had the blip on his scope and was positively identified.

The old method of transferring traffic from one facility to another was very dangerous, as the accident proved. Instead of giving a positive radar handoff, the aircraft were released from one controller to another by ETAs (estimate time of arrival) alone. Airline passengers are well aware of the many variables that can cause a flight to be late. Human calculations, weather, and winds are only a few factors that go into the makeup of an ETA. Yet this criterion was used to control traffic. Airplanes were shuffled from one control to another—strictly on ETAs. There were precious minutes when planes flew in crowded conditions without the eye-in-the-sky radar.

Stop signs are never erected until a costly accident. The FAA once again waited until the inevitable happened.

The mid-airs were on the wane and the old feeling of "Don't talk about it, we haven't had any in years," prevailed. But John Q. Public was once again shown that all was not well in aviation. On December 4, 1965, an Eastern Connie and a TWA 707 collided over Carmel, New York. Due to the skill of both pilots, only four people lost their lives. The 707 limped into JFK with part of its wing gone.

The commander of the Eastern Connie, Captain White, did one of the most spectacular pieces of flying I ever heard of. With literally all control to the tail section gone, he managed, by uncanny use of his throttles, to make a crash landing. Unfortunately, Captain White lost his life after landing when he reentered the burning plane to rescue a passenger.

With the help of a few politicians and the daily bombardment of editorials in the *New York Times,* the FAA was once again on the spot and, on November 9, 1967, the floor of the positive control was lowered from 24,000 to 18,000 feet. This positive control would only effect certain parts of the northeastern United States. Now all aircraft flying in this area above 18,000 feet must be on an IFR flight plan, and under positive control on or off airways. Because of this improvement, the "anything goes" airspace was lessened.

For some time the FAA had been experimenting with a radar system called Alpha Numerics that would give the controller three dimensions. Before this, a radar man could not tell the altitude of his traffic, only the direction. This system was installed in the Indianapolis center in order to evaluate it. The men who worked with it had numerous complaints, some of which follow:

1. It took too much time for the coordinator to set up the system. At least 15 separate buttons per aircraft had to be pushed in order to feed pertinent information to it. While this process was being carried out, the traffic team would be short one man.
2. There was so much information displayed on the scope that the primary target disappeared under the maze (flight number, altitude, destination, beacon codes, etc.).
3. Many times, when two targets would merge, the data from one would jump to the other and vice versa.
4. Only the coordinator would view the scope from the front. Both radar men, seated to each side of it, had to read their data sideways.
5. It had a great tendency to drift off target.

The men at Indianapolis center would not use the Alpha Numerics when traffic became heavy, and this was the very time when it was needed most. This, then, was the fantastic cureall that would appease public pressure and make mid-airs a thing of the past.

In February 1967 the Alpha Numerics was bid a fond farewell by the men of Indianapolis center. With a "So long, we're glad to see you go," they applauded while Alpha Numerics was moved and installed in the center at Islip, New York, at a cost of over $2 million. The idiosyncracies of the system were well known to the men in New York center. Behind some of the panels were inscribed such statements as "We couldn't make it work maybe you can" and "If you can use this with more than three targets you're an ace." And in one of the ash trays was the inscription: "You have just used a million dollar ash tray."

Alpha Numerics received the same unfriendly treatment in New York that it got in Indianapolis. The Professional Air Traffic Controller Organization (PATCO) told its members not to use it as it was unproved and should not be utilized to control the lives of the traveling public. In 1968 Alpha Numerics was quietly and unceremoniously dismantled and shipped off to Atlanta, Georgia. The FAA hopes it has found a permanent, happy home.

On September 9, 1969, student pilot Robert Carey arrived at Indianapolis' Brookside Airport and began preparing for a solo cross-country flight. With a total of thirty hours in the air, he needed only ten more before he would be eligible for his license.

He was briefed on the weather by Indianapolis flight service personnel who said it would be a good day for his trip. A VFR flight plan was completed and filed. (A VFR flight plan is useful only when a flight is overdue: rescuers would have an idea where to search. No one in ATC is advised of the flight, the route, or the altitude.) Carey's trip was one of the requirements for a student pilot before he can be issued a private license. After preflighting his single-engine Piper Cherokee, he climbed aboard and fired up the engine. At 3:21 P.M. he took off and, after following the prescribed departure procedure, turned on course. In eight minutes he would be dead.

Captain James Matthew Elrod, a senior pilot with Allegheny

Airlines, was in command of Flight 853, a DC-9 from Boston to St. Louis. Flight 853 was operating under IFR and had just departed Cincinnati, its last scheduled stop.

Allegheny is a good line to fly for. Leslie O. Barnes is probably one of the best-liked presidents of any airline, and he has run Allegheny practically from the start. His pilots know that if they ask for any safety features they will get them. Allegheny's fleet of planes are equipped far in excess of FAA requirements: items such as autopilots, strobe lights, and radio altimeters, to mention a few. It takes more than this to beat the unsafe system, however.

Indianapolis approach control instructed the Allegheny flight to turn to a heading of 280 degrees and descend from 6,000 to 2,500 feet. At 3:29 the Allegheny target disappeared from the radar scope. The two planes hit and spun to the ground; all eighty-two on board perished.

The ceiling at the time of the accident was reported by two airline pilots at 3,500 feet. Pilot Carey was cruising at 3,500 feet, which would put him directly under the cloud deck. FAA regulations state that a VFR pilot must maintain at least 500 feet below the clouds. Not a very safe margin of time for Allegheny to spot any traffic as he popped out of the overcast—but all perfectly legal.

On September 10, the day after the tragedy, FAA Chief Schaffer and his deputy administrator, D. D. Thomas, were questioned by Congressmen Dingell and Ottinger concerning the catastrophe.

"I choose to think that yesterday's accident was fate," said Schaffer.

Congressman Dingell summarized the horrified reactions of all present: "I am much troubled when I hear an official of the FAA come up and tell me that some eighty people died because of a stroke of fate."

Congressman Ottinger added: "I think it is time that the FAA did have some biases and prejudices in favor of the safety of the people in the air. . . . Mr. Thomas, I think you are very largely responsible for it. I would think those eighty-two deaths are very much on your conscience because I think you, the FAA as an agency, could have done something about this, and ought to be scrambling to do something about it today."

Mr. Schaffer had this to say: "If there were a contributing factor, it was simply due to the limitation of radar, and the lack of a transponder on the light plane."

Months before the accident, the radar at Indianapolis was not working properly. Controllers had made repeated writeups and had warned their superiors about the numerous problems with the radar. The FAA official supervisor's log, in which daily radar discrepancies are noted, read as follows:

Sept. 2:	1239 Radar out of service
	1244 Radar resumed operation
	1220 Radar on low power
	1652 Radar normal
Sept. 4:	0938 Amplitron out of service
	0950 Amplitron normal
Sept. 6:	1524 Radar on low power
Sept. 7:	1524 Radar still on low power

On September 8, the day before the accident, the log revealed the following.

Sept. 8:	1620 Radar on low power (been on for three days)
	0037 Radar on high power
	0038 Radar on low power

Indianapolis radar operates on low power (magnetron) most of the time. The equipment is so old and rundown that operating it at normal power overloads the system. Once the power is reduced, the possibility of spotting primary targets (non-transponder-equipped aircraft) is nil. When the center switched to low power they reduced their traffic separating capabilities from 2,800 kilowatts to 355 kilowatts, which gives approximately 12 percent of the potential power output.

After every accident of noted malfunction concerning radar, the FAA sends a flight-check ship to report on the disposition of the equipment. During the months of April and July flight checks were made of the Indianapolis radar system. Both checks disclosed numerous areas that did not show targets. The PATCO check revealed that no effort was made in either instance to replace parts or to peak the system (adjustments made to increase target returns). The checks were made simply to appease the numerous squawks of the controllers. Experts pointed out that the oscillator tube of this radar had been one of the biggest problems. This tube had been replaced on an average of once a month for the last six months. The tube was of such importance that if it were to drift off tolerance there would be wedge-shaped portions on the scope where traffic would be invisible.

Jim Hays, president of PATCO, stated that shortly after the collision, FAA management ordered new parts installed in the Indianapolis tower radar prior to the official check of the system. Technicians swarmed over the electronic equipment adjusting it to its peak power. The tests were run and the equipment was given a clean bill of health. The Indianapolis radar passed the tests and then was allowed to deteriorate. The day after the tragedy, another Allegheny's Flight 853 narrowly missed hitting another light plane.

Controller James C. Knecht, a radar man at Indianapolis tower, was questioned by the press concerning the equipment he had to work with. He told them "The radar was a hazard to life." On September 13, 1969, Mr. Knecht was called to a meeting with the chief controller of the Indianapolis tower. Knecht was told: "You will never work radar in this tower again." Controller Knecht was given menial tasks to perform and subsequently fired. He had told the truth about the system.

Indianapolis radar is not the only system that has bugs in it.

Much of the radar that guides planes is World War II vintage. Most of the radar antennas used by the FAA were never intended for traffic-separation use. The antennas were primarily built for the nation's defense and for use by the Air Defense Command (ADC). The tilt of an antenna used for spotting enemy aircraft hundreds of miles away and guiding fighters to them is not to be considered accurate for other purposes. But the FAA simply taps into the military antennas and uses them for their own operation.

Since the antennas are sometimes hundreds of miles away from the traffic controller, a way must be found to bring the pulse from the antenna to the radarscope. This is done by a series of microwave Link towers spaced 30 miles apart. The radar signal is passed from one tower to the next until reaching the scope of the controller.

What happens when one of these towers fails? At nine o'clock in the evening, November 29, 1969, all JFK departure traffic stopped. It remained stopped for over five hours. Pranksters had climbed one of the Link towers at East Meadow, New York, and had broken the antenna. These signals should be transmitted by existing telephone lines; this would be more realistic and would allow for a backup.

Since the microwave towers are above ground, they create other potential dangers. A fast-moving cold front can shift a beam between towers and can wipe out returns of the scope completely.

The FAA allows a 3 percent error in the alignment of their antennas. This error is small near the antenna, but at a distance of 100 miles, two antennas could be off by as much as 3 miles each. The controller who receives a handoff from another sector would have his traffic show up 6 miles from where it should be.

A senior maintenance technician at New York center told me some of the dangerous practices at a center that works traffic in five states. When a controller at the scope reports a malfunction with his equipment he immediately checks with maintenance, down in the bowels of the center. Now the technicians have to move rapidly; hundreds of lines depend on an accurate radar right now. Each scope is supposed to have a maintenance display monitor to allow for a quick check and remedy; 40 percent of these necessary units were taken out and given to another facility.

New York center needed five new RBDE-5 display units. Rather than spend the money for the new scopes, five old RBDE-4 ones were shipped in. These units were direct mismatches and would not work when mated to the units that supply power to them. Since all equipment must pass a JAIT (joint acceptance inspection test), special power supplies were used to pass the test. On completion of the check these power supplies were put back on the delivery truck and returned to the depot. Now the center has five spare display units that are completely worthless.

One of the junction boxes (a large unit that contains numerous wires that can be quickly plugged in or changed, similar to a switchboard) was a potential hazard. The patch panel for rapid replacement of scan and beacon control had wires of such length that they were lying coiled on the floor of an extremely narrow walkway. A person passing this unit could easily disconnect one of these plugs, thereby shutting off the power supply to a vital unit.

Traffic at JFK was backed up six hours because of a breakdown in the system. During the peak rush hours the approach radar controller noticed that his scope was out of "synch." Because of a faulty gearbox bearing, the radar continued to scan the sky but made targets appear south of the field, when in reality they were north. While incoming trips were stacked up as far as 200 miles from New York, a check revealed that JFK lacked the replacement parts. A spare was found "on the shelf" at Newark Airport. It took five aggravating hours before the faulty bearing was replaced and the traffic started to flow again. Two days later, Kennedy radar failed again.

In August 1968 the New York common IFR room opened

for business. The brainchild of the FAA's D. D. Thomas, this facility was to be the answer to the traffic confusion in the New York area. Instead of the three major airports (JFK, LaGuardia, Newark) having their own separate approach and departure controls they would be combined under one roof, at JFK. A special wall-type radar display was developed. The main antenna through cables and electronic boxes fed a signal to a projector. The display was carried to two large wall screens by means of a high-powered bulb (similar to a slide-projector lamp). This lamp has the same infuriating characteristics as the one that blows at home just when everyone is ready to view your vacation pictures. It burns out two or three times a week. Each control position in the common IFR room also has the old-fashioned small display. Controllers have learned not to use the wall screens at all. At a cost of $150,000 apiece, they remain: a pair of highly priced modern murals.

The antenna wiring that supplies signals for both LaGuardia and Kennedy radar is underground; Newark still depends on microwave towers. All of Newark's problems are not with the towers alone. The full complement of men for this position is ten: seven journeymen and three trainees. In order for a man to eat lunch, another controller works two positions.

Working at a radarscope is ticklish business when the scope is viewed as it was designed, from the bottom. But there isn't enough space for the Newark handoff man to sit and view the scope at the bottom so he does his work sitting on the top where he must view and mark targets upside down. Since the radar men have been taken out of the individual towers, the only way they can receive the weather information is from two wall screens. At least these displays don't blow bulbs, but because of the distance from the controllers' position to the wall, many times their view is blocked by the heads of other men.

Controllers use small strips of cardboard to write pertinent information concerning the many flights they are guiding. These flight strips are placed in metal holders which in turn are put in racks at the panel. The equipment panels were not made to take these strips, so small wooden boxes were constructed. These makeshift boxes do not hold the strips properly, thereby increasing the workload again.

On January 9, 1971, American Airlines Flight 30 was approaching Newark Airport after a nonstop flight from California. The weather was a measured ceiling of 3,300 feet and the flight visibility a scant three-quarters of a mile.

The traffic controller directed American: "Turn right heading 180 degrees." One minute later the radar man called again: "American 30, traffic at 12 o'clock less than a mile, northeast bound, slow."

American replied: "No contact."

Inside the cockpit the captain and the first officer were busy with their check list, not gazing out the window.

"Everything is murky up here."

"Boy! it is, and I suppose it's VFR."

"Well, another thousand down is, but I hope nobody is—

"Damn."

"What's he doing up here?"

American's frantic call to approach control "We have been hit by that airplane, American thirty."

The American crew were able to land at Newark in spite of major damage; the Cessna 150 that rammed them crashed. Both student and instructor died. The jet crew reported that they had a brief glimpse of the Cessna just before the mid-air but there was insufficient time for evasive action.

Another shining example of the FAA system in operation. Both planes were as legal as could be. The jet was on an IFR flight plan, the Cessna had no flight plan and was maintaining VFR. No radio contact was made to the Newark approach controller, *nor was one required*.

On June 6, 1971, an Air West DC-9 was climbing to the IFR assigned altitude of 33,000 feet. An F-4 marine jet fighter was also climbing up over Duarte, California. The Phantom pilot was climbing to 15,000 feet on a VFR flight plan. As he reached altitude he decided to roll his plane for kicks. A few moments after he leveled out the radar observer in the back seat looked up in horror to see the fuselage and left wing of the DC-9. The fighter pilot rolled again to the left and rammed into Air West. The only person to escape this disaster was the radar observer.

The marine Phantom jet was not in contact with the center and his transponder was not operating. His flight plan did not call for radio transmissions with the center controlling the airliner and his inoperative transponder made it impossible for the traffic controller to pick him up on his scope and warn the airliner. Another fifty people dead.

It was great flying weather and Sergeant Yoshimi Ichikawa, twenty-two, was on a training flight in his F-86 fighter. Flying on his wing in a similar plane was his instructor. Student pilot Yoshimi had only 21 hours in this aircraft. As the two fighters were leisurely cruising over the main Japanese island of Honshu, suddenly Yoshimi's earphones came to life with the screaming of his instructor pilot:

"Climb and turn right! Climb and turn right!"

Too late. With an earsplitting crash Yoshimi's jet rammed an All Nippon Airways Boeing 727 and 162 bodies and debris were scattered over a 20-mile area. This was the world's worst air disaster and it was a mid-air.

On August 14, 1971, just south of Los Angeles, a Cessna 150 and a Continental Airlines 707 brushed against each other. This time disaster was averted. The 707 landed at Los Angeles and ninety-six people were glad to be alive. The two occupants of the light plane escaped death as they crash landed in a city dump.

On December 5, 1971, at Raleigh, North Carolina, a most bizarre scene took place. Garey Solloman, a ground crewman for Eastern Air Lines, was busy at work when an unbelievable sight caught his eye. Eastern Flight 898, a DC-9, was approaching the field for a scheduled landing, only this time the airliner was flying directly over the center of the airport, and something was hanging underneath it.

"They were stuck right together coming over the runway," said Solloman. "We could see pieces flying off and then the little plane just dropped off and fell to the ground." What Garey Solloman had seen were the remains of a single-engine Cessna and its two occupants as they hit the ground. The light plane collided with the airliner on final approach and was carried for nearly a mile locked in the airliner's landing gear. Over the airport the Eastern plane, like a giant eagle tired of carrying its prey, suddenly relinquished it. After circling the field for forty-five minutes to burn off fuel and have the landing gear checked, Eastern Flight 898 landed and twenty-seven lucky people had beaten the system.

At any time during the daylight hours, approximately 10,000 planes are in the air over the United States, 80 percent of them flying under the "see and be seen" rules of VFR *and authorized to do so below 18,000 feet*. The other 20 percent are on IFR clearances. The National Aviation System Policy Summary, Department of Transportation, for March 1970 shows an interesting graph. They forecast that with no change in the regulatory policy, by 1975 there will be five to ten mid-air collisions between air carrier and general aviation aircraft per year.

Former PATCO president Jim Hays made this statement after the Allegheny DC-9 crash: "There are going to be other tragic collisions in airport areas in the near and immediate future unless steps are taken right now to alleviate the deficiencies in the system. There should be no further delays. The FAA should not try to camouflage as it has, not only in the Indianapolis crash but also in San Juan and others, covering up its own deficiencies and shortcomings while the traveling public's lives are at stake."

Mr. Hays continued: "If the eighty-three persons who were killed earlier this month realized the deficiencies in the radar system and that the FAA considered the system a 'reasonable margin of safety,' they would not have flown.

"The reasonable margin of safety is the philosophy of the FAA officials in their entire approach to the airport and airways system. It must be changed before more people meet their death."

"The greatest threat to loss of life is still the mid-air collision. It is likely that the probability for a mid-air accident of major magnitude will increase unless the FAA takes positive action to correct problems and decelerate the implementation of a new automated control system known as RDP (radar data processing)." These are the words of John B. Galipault, president of ASI (Aviation Safety Institute).

ASI is a non-profit research organization able to approach safety from all perspectives. Mr. Galipault has taken up the long line of slack that developed when the FAA discontinued its immunity program in 1971. John has sent out thousands of anonymous reporting forms to controllers and pilots so that he can get a broad picture of the inadequacies of the system.

Since January 1974 ASI has received an average of forty-four controller reports per day. This reflects only about 4 percent of the number that could be received. The following are a few of the terminals and the number of their discrepancies for a two-month period:

Amarillo, Texas	*31*
Denver, Colorado	*37*
Guam	*142*
White Plains, New York	*28*
Dulles, Virginia	*32*
Kennedy, New York	*15*
Rochester, New York	*25*
Louisville, Kentucky	*181*
Toledo, Ohio	*25*

There were many more too numerous to mention individually. But during the two-month period in 1974 there were a total of 1,152 reports filed with ASI.

The FAA is presently revamping its air traffic control with a new radar system that has the potential to cause rather than prevent mid-air collisions. This system should have been tested more completely before it was put to use on the unsuspecting air traveler. Perhaps the FAA would rather deal in averages. Not too many student controllers run two planes into each other while they are learning. Perhaps they feel that the radar will also "luck out."

The big difference between the old system and the new RDP is as follows: The older wide-band radar antenna picked up a target and it was displayed directly on the controller's scopes; the new RDP signals received from targets by the radar antenna are routed to a digital computer. The computer picks up the appropriate target and projects it as symbols on a video display. In essence the target that the controller tracks is not the actual aircraft he is controlling but rather a display put there by the computer. If the computer is not set properly, the targets will be depicted many miles from the actual aircraft.

RDP is a poor instrument when tracking aircraft without a transponder. The older wide-band radar could display these targets most of the time, but with RDP it is practically impossible. Weather display is almost non-existent on narrow band as compared to wide band. This is a most important feature. When a pilot requests a turn to escape a thunderstorm, the controller is at a loss because he doesn't have the precipitation area on his scope.

In order to operate the RDP, a coantroller must push a multitude of buttons. Each aircraft that a controller accepts requires pushing a minimum of thirty-six buttons. Not a demanding task during light traffic, but when a man is working ten aircraft, 360 buttons must be depressed.

Between February 20 and March 20, 1974, there were no less than eleven near misses directly attributed to RDP ambiguities. The data block associated with each target shows the flight number, altitude, transponder code, etc. This is all vital information, but when mismatched, information is displayed on the wrong target, or when data blocks bunch together, it becomes impossible to read them.

So now we have an untried system being worked by student controllers which doesn't make flying as rosy as the airlines would have us believe. It is especially frightening in view of the fact that the skies are getting more and more cluttered with aircraft. The following chart indicates the growth of air traffic by projecting the number of planes that will be in the air:

	1975	*1980*
Airlines	*2,577*	*3,019*
Military	*20,323*	*20,119*
General aviation	*158,000*	*186,000*

In 1974 the general aviation manufacturers delivered more than 14,000 new aircraft, and they forecast delivery of over 15,000 this year.

On May 21, 1973, a North Central Airlines and an Air Wisconsin DHC-6 collided over Lake Winnebago, Wisconsin. All on board were killed.

On October 11, 1974, four occupants of a light plane were killed when they were rammed from behind by an Air National Guard F-106. There were two F-106 Fighters running intercept problems. One of them locked onto the wrong target and four civilians died.

One month later on November 12, two USAF F-111A fighters were up on a refueling mission. There was a malfunction in lead's navigation instruments, so he signaled his wing man to take over. During the confusion the new leader saw a flashing red light that he believed to be the tanker. It wasn't. It was a private Aero Commander which he flew through.

In 1973 there were twenty-four mid-air collisions in the United States. In 1974 the number jumped to thirty-two.

These collisions will continue as long as the rules are not changed. The "see and be seen" concept should be abandoned for airliners. They should be in a positive airspace from takeoff to landing. The mixture of VFR and IFR traffic has never worked and it never will.

Chapter Twelve

WINDY WASHINGTON

On February 10, 1956, President Dwight D. Eisenhower appointed Edward P. Curtis his special assistant for Aviation Facilities Planning. On May 10 of that year Curtis submitted his report to the president. In part the report warned: "A crisis is in the making as a result of the inability of our airspace management system to cope with complex patterns of civil and military traffic that fill the sky. The growing congestion of airspace is inhibiting defense and retarding the progress of commerce."

On March 8, 1961, President John F. Kennedy requested FAA Administrator Najeeb Halaby "to conduct a scientific engineering review of our aviation facilities related research and development and to prepare a practical long-range plane to insure efficient and safe control of all air traffic within the United States."

In November 1961 FAA Administrator Halaby endorsed a new and unique method of transporting passengers from the terminal area at Dulles International Airport to their aircraft. Twenty special mobile lounges were built at a cost of $4,654,660. (This would have paid for a lot of new radar.) Dulles is one of the finest airports in America, but because of its distance from Washington it is one of the FAA's costliest white elephants. Washington National Airport handles double the traffic of Dulles, and without mobile lounges.

On July 18, 1966, the FAA's Intercom Bulletin reprinted a letter from President Lyndon Johnson to "Dear Bozo" (General William F. McKee, FAA Administrator):

> I have noted with satisfaction the excellent work which you and your associates at the Federal Aviation Agency have been doing in reducing costs and manpower while absorbing additional workload and improving services to the public.
>
> I have taken particular note of your cost reduction program under which you saved $47 million during the 1966 fiscal year. These savings have been accompanied by a reduction in Agency employment more than 3,500 employees—eight percent since 1963. The Agency has succeeded in combining economy with a safety program which has helped the commercial air carriers of the United States achieve the best safety record in the world and the best record for any five-year period in the history of American aviation. You have clearly demonstrated that these goals are in addition to the more than $100 million of savings accomplished since 1960.

In October 1967 the air traffic controllers held their convention in St. Paul, Minnesota. One of the principal speakers was General McKee. Here are some excerpts from his speech:

> I am determined to know more about the conditions under which you work, and finally, I am determined to do whatever is necessary to improve these conditions. We don't have enough controllers. [The $47 million he saved would have paid for the training of 2,300 controllers.] This insufficiency has meant many of you men at the consoles [radar] have had to work overtime. It has meant many of you gave up your vacations. Many of you had to cut weekends short and some had to come back on short notice during off hours. There will be a need yet for overtime.

> There will be a need for some six-day weeks . . . a need for sacrifice. . . .
>
> My job is to provide you with the tools and personnel to do the jobs. I am your man in Washington. I welcome, consequently, every opportunity to champion your cause.

It seems that the story "Bozo" tells his subordinates and the one he tells his superiors differ considerably.

On July 8, 1969, the Air Line Pilots' Safety Forum was held in Atlanta, Georgia. The present FAA chief, John H. Schaffer, had this to say:

> The air traffic situation in this country has reached the stage where the existing tried-and-true system is strained to the utmost. It is too frequently supersaturated in some areas. Revenue passenger miles will triple in the next ten years and general aviation traffic will probably grow to at least twice current levels in the same time.

When Richard Nixon was campaigning for the presidency he made the following statements concerning national air policy.

> A first priority of my Administration will be to strengthen our air controllers' force, improve their working conditions, and provide them with new equipment they need to keep our airways safe.
>
> Our traffic controller force which has not been increased for four years despite a 60% increase in their workload, works long hours under tension-filled conditions with inadequate pay and retirement benefits.
>
> Today, however, years of neglect at the highest levels of government have produced a crisis in air transportation which requires new and imaginative solutions.

The next chapter will demonstrate the crying need for drastic air traffic improvements *now*.

Chapter Thirteen

THE GREAT CONTROLLER

MOST AIRLINE passengers are well aware that it takes mechanics to repair the planes, pilots to fly them, and towers to issue traffic instructions. Very few are aware of the hundreds of people concerned with behind-the-scenes control of air traffic.

In July 1927 the first coast-to-coast air service was inaugurated, extending 2,612 miles. Since all flying was done in good weather only, the airways were marked by flashing beacon lights. The government placed these facilities under the Bureau of Lighthouses. As aviation grew and instrument flying techniques developed, flight during inclement weather became a reality.

In December 1935 the first air traffic control center came into being at Newark, New Jersey. It was organized and manned by private airlines. Its purpose was to provide information to airline pilots concerning the whereabouts of possible conflicting flights. Even in the early days of flying, the government had to be prodded, and it was not until seven months after the first privately operated center that the Bureau of Air Commerce took over operation of the then existing centers at Newark, Chicago, and Cleveland. Today there are twenty-seven centers (each center controls an area from two to four states) that guide traffic along the 350,000 miles of federal airways.

Here is a brief description of how your pilot gets issued his route. All scheduled airlines have prestored flight plans in the various centers. If a route change is requested by the pilot, it must be called into the center. While you, the passenger, are trying to figure out which side of the plane will afford the most shade, the men up front are busy with the flight plan. A radio call is made to the local control tower and the clearance is put "on request." The flight data man at the tower phones the controlling center and asks for appropriate clearance. Traffic information en route has been thoroughly checked and clearance is issued to the tower. On receipt of his clearance the pilot copies it down verbatim and, as a check, reads it back. In essence the tower controls all traffic on and about the airport. And you are under the watchful eyes of center radar from approximately 20 miles from your departure airport to an equal distance from your destination.

In November 1968 the FAA's training academy in Oklahoma City, after having been closed for almost three years, reopened its doors to controller training. The plan was to train approximately 2,000 controllers per year for the next five years. A definite step in the right direction, but only a drop in the bucket. The FAA states that they presently have 24,000 air traffic controllers. It is true that they have many men in various control positions, but only 8,400 are journeymen controllers. Over double that many are urgently needed now to support the existing system, not to mention future expansion. The FAA hired 2,428 trainees from January through April 1971; during that time 1,288 failed to make the program.

Even when a controller graduates he does not have his license. This he gets after years of training at the various traffic control facilities. The practice of allowing trainees to control live traffic should be abolished. Pilots are not taught to fly the Boeing 747 with three hundred passengers on board. Why should inexperienced controllers be responsible for human lives?

The training academy is understaffed and overcrowded. Most of the instructors spend eight hours a day teaching. This is a tremendous burden because class hours are only a small portion of the time necessary to prepare and grade exams. On December 14, 1969 the FAA lowered the minimum age of controllers from twenty-one to eighteen; now teen-agers will be responsible for hundreds of lives.

Students are on a two-shift basis with classes running from 6 A.M. to midnight. A prospective candidate is actually hired at a regional office near his home. He is given the necessary tests, and if he is accepted, he is sent to Oklahoma City. He receives nine weeks of schooling and is then sent back to the region that hired him. It is interesting to note that the academy instructors lack the power to hire or fire. They can only recommend. If a candidate "flunks out," he may be hired at the facility where he originally applied. The academy has a 30 percent washout rate.

Training aids at the school are ancient. Student tower operators still push model planes around a mockup airport painted on the floor. Radar trainees are schooled on sham radarscopes. The instructor can flash blips on a screen, but any resemblance to working radar is purely coincidental. The flight data boards are the only aids that see good replicas, and that is mainly because they are inexpensive. Much of the training is done by means of a simple 23-inch TV screen, set up before a class of twenty students. One comment about personnel shortages: "In en route training we are 8 teachers short. Our total setup is meant for 160 students at a time while we are really training 240 in the en route portion alone."

A graduate student is barely able to find the right manual to look up a regulation. Since the accent is on speed and on cranking out controllers as rapidly as possible, the brunt of the workload falls squarely on the shoulders of the journeyman controllers at the various facilities.

The average time it takes to be a full controller is three

years. This makes nearly every air traffic control facility in the United States a training school. Not a very comforting thought. Now the already overworked journeyman has to keep one eye on the traffic, and the other on his student. (More on this later in the chapter.) Present training rates will just keep up with normal attrition. Many of those who entered air traffic service in 1946 are reaching the burnout age of forty-five. There should be an early retirement for this highly specialized job, but as of this date there is none in sight.

For years, the job of keeping airplanes from running into one another has been a nerve-wracking one. The men get little recognition for their work and have deplorable operating conditions. Two trailblazers decided to do something about it. On January 7, 1968, Mike Rock, a controller from New York's LaGuardia tower, and Jack Maher, from the New York center at Islip, joined forces and held the first meeting of what was later to be the Professional Air Traffic Controller Organization (PATCO). In the summer of 1968 the PATCO controllers staged a safety campaign. Some called it a slowdown. I prefer to think of it as "flying by the book" and not on the hairy edge of disaster.

A controller is indeed a special individual. He must make rapid decisions and they must be right. He must keep track of from six to twenty planes at one time. Peak-hour traffic at, say, Miami goes as high as 120 airplanes (5,000) passengers per man-hour. Even with a small amount of traffic he still controls the lives of at least six-hundred people. On busy days he may not leave his sector for eight to ten hours—eating a sandwich while guiding planes through the maze of skyways.

The JFK common IFR room is truly the nerve center of New York. Here all instrument landings and takeoffs are monitored. Airplanes fly over and under each other in unbelievable patterns. Departures turn over arrivals and tunnel under the corridors of international flights. Only a few degrees off course or a small variance in altitude can produce disaster.

Dr. Wayne Sands' fact-finding medical report, sanctioned by PATCO and read before Senator Vance Hartke in connection with retirement benefits for government employees, had the following to say:

> One controller in a coma after a serious operation continuously talked to imaginary pilots in airplanes. He could only be quieted by a colleague who acted as his radar handoff man. The trouble with the air control business is that they find it difficult to forget their job even when they are off duty. . . .
>
> A top-notch controller in his early thirties was working his radio during a crash in which many lives were lost, including the pilot to whom the controller had last spoken. The controller became irritable, sleepless, and withdrew from his friendships and sports activities. He had nightmares in which the pilot blamed him for the mishap. An investigating board exonerated the controller, but he could not acquit himself and became acutely anxious and "felt as if he were climbing the walls." Hospitalization, medication were needed. He returned to work for several weeks but an air tragedy in another country reopened his psychological wound. On the job he became terrified and his blood pressure and heart rate were greatly increased. He was hospitalized again, and never returned to air traffic control.

There are cases on record where working controllers have had fatal heart attacks caused by stress.

An FAA task force report had this to say about controllers: "Office of Aviation Medicine found that certain stress related to diseases were one and a half to four times higher in air traffic employees than for the general 'airman' population. It is up to eight times higher than is usual for USAF pilots."

I asked three PATCO officers, Mike Rock, James Hays, and Ed Janata, for their thoughts concerning the traffic crisis and tape-recorded their answers.

REPORTER: Can you cite any specific cases where a supervisor has pressured a man?

HAYS: I was working approach control position [Center radar hands off traffic to approach control. He in turn hands it off to the local controller] at Chi-

> cago O'Hare. Local asked for a minimum of five miles separation to touchdown. I ran six miles to allow for the fact that when the traffic gets by the outer marker [approximately five miles from the field] they slow down about thirty knots. If I hadn't allowed an extra mile the traffic wouldn't have kept the proper separation. A man from the area FAA office stood behind me for about fifteen minutes. He left and the watch chief came over and observed my operation. He asked me why so much separation and I explained it to him. I was removed from the position, reprimanded, and received a letter of discussion. . . . A Denver controller was suspended for fifteen days because he requested help from another controller.

In Atlanta a controller was told to work two positions. He said he could not control the additional work. He got ten days off.

REPORTER: Mr. Janata, do you have any cases from Miami tower that you would like to relate?

JANATA: Jacksonville center requested twenty miles' separation from us. And rather than let me hold my traffic for a minute or so I was told to roll them. Supervisors don't like to have a flow control restriction [traffic prevented from taking off hundreds of miles away as there is no room for them on the airways] entered in the log during their shift. They think it looks as if they don't know how to cope with the traffic. So I was told to give the traffic a normal three miles' separation and build in the other sixteen. This of course I can do, but it will increase the workload for myself and the pilot. Instead of proceeding on course I have to issue rectors [turns] in order to increase the interval.

HAYS: On August 20, 1969, on the midnight shift at Denver center, a watch supervisor ordered a trainee with only one week of training to control high-altitude jet traffic for over four hours with no instructor to monitor his judgment or provide guidance.

REPORTER: Is there any specified number of planes a man can handle at one time?

ROCK: A survey run by the FAA stated that a man should be able to handle eight at once. Any more and safety would be compromised.

HAYS: At O'Hare airport peak IFR traffic reached 220 in one hour. The controllers were commended by their chiefs.

ROCK: I ran 118 operations in one hour at LaGuardia with 5,000-foot crossing runways. There wasn't any way in hell that a man could do this legally. But if you want to eat, you do it.

HAYS: I have never seen a man reprimanded for running traffic too close. They have a saying in Chicago: three miles is maximum—there's no minimum as long as they don't hit each other.

JANATA: When the FAA flight checks a navigation aide they usually manage to do it during periods of peak traffic. They fly down the runway against traffic and in general cause added workloads. These checks should be performed at night or early morning. FAA claims they can't afford the overtime. I can remember one specific case when the check ship was making runs perpendicular to traffic. I stopped all inbound and I was told by my boss that flight check was VFR, there was no problem. Well, I saw plenty of problem because when a check ship is running recorded approaches those pilots are so busy they don't have time to look out and keep themselves clear.

REPORTER: Do you have adequate backup equipment?

ROCK: Backup equipment is just about nil.

REPORTER: What happens when the radar or radio goes out?

ROCK: Chaos. I wish I had a dollar for every time the radar has quit. I have seen radar fail in two or three sectors at one time. This means that approximately a hundred planes are in danger of colliding. The only course that the pilots have to take is to try and separate themselves. It's a frightening situation. About all I can do in a case like this is pray the Great Controller will do his job.

REPORTER: Who is the great controller?

ROCK: God. . . . Economy is the watchword in the FAA —cut corners, cut expense, move the traffic, and don't rock the boat. You know it is really a sad situation with the FAA and money. I know for a fact that the FAA spent $60,000 to pipe music in their Western Region offices.

REPORTER: Mr. Rock, do you know of any controller who shouldn't be one?

ROCK: I certainly do. Take the man who has worked light traffic in Podunk for fifteen years. He decides he wants to get more pay and work in a higher density position. Simply because we are short of men they will check him out and stick him on the position. He has no more business there than the man in the moon.

REPORTER: Have you ever seen a man panic on the position?

ROCK: There is one vivid experience I will never forget. A man froze on the mike. His eyes rolled back in his head and he broke out in a cold sweat. He keyed his microphone continuously but said nothing. A nervous breakdown right on duty.

REPORTER: Is there any regulation that specifies just how many positions one man can work?

ROCK: If they could cut me in eight pieces they'd have me working eight positions.

REPORTER: Are there many trainees in the facilities today?

HAYS: Well, the O'Hare tower has slots for eighty-five qualified controllers. They have ten trainees. The rest are checked out.

ROCK: Trainees should not be allowed to handle line traffic. They should be qualified at the academy.

On March 5, 1969, Prinair Flight 277, a four-engine airliner, was en route from St. Thomas to San Juan, Puerto Rico. The weather was good, with a 2,000-foot overcast and scattered areas of precipitation. The departure of Prinair was handled by trainee Carl Mandrell under the watchful eye of instructor Joe Cioffera. Flight 277 requested an altitude of 3,000 to Fajardo intersection which is located 20 miles east of San Juan VOR. Mandrell declined this request because another Prinair flight was already at 3,000. Four thousand feet was the level assigned and acknowledged by Prinair. Seven miles east of Fajardo intersection, San Juan approach control received a handoff from center and took over guiding the flight. At San Juan, trainee Gerald Belanger was manning the approach control position and was in turn being observed by instructor Fred Read.

PRINAIR: San Juan approach control, this is Prinair Flight 277 level at 4,000.

BELANGER: Prinair 277, San Juan approach control radar contact, three miles east of Verdes fly a heading of 250 for a vector to ILS final maintain 4,000.

PRINAIR: Roger, fly heading 250 for vector to the ILS and maintain what?

BELANGER: Prinair 277, maintain 4,000.

PRINAIR: OK. We'll maintain 4,000 turning to a heading of 250.

At this time the traffic coordinator was relieved of his position by the supervisor and directed to assume another position in the tower. Instructor Read was saddled with two important positions. He now had the task of watching his student and at the same time receiving handoffs from the center. The spotty precipitation to the south was becoming more intense and starting to saturate the vectoring course. Earlier, Belanger had guided numerous transponder targets through this area with no problems at all.

BELANGER: Prinair 277, San Juan approach turn left heading 220 descend to and maintain 3 for vectors to the ILS.

PRINAIR: Roger. Maintain 3.

While involved with other responsibilities Read and Belanger followed Flight 277's blip to a position east of the precipitation area. It was at this position that Prinair asked for and was refused a lower altitude.

Instructor Read became fully occupied with his duties as handoff man and left Belanger alone at his scope. Since Belanger was also controlling five other aircraft he could not possibly give Prinair his undivided attention. When instructor Read next looked at the approach radarscope he did not see Flight 277.

READ: Where's 277?

BELANGER: He's somewhere in the precip.

READ: Find him quick. [Read again returned to coordinating duties. Belanger observed two targets in the general vicinity of where Prinair should be. He spoke in an anxious voice.]

BELANGER: Prinair 277, turn back right now to a heading of 280. Prinair 277, this is San Juan approach. Turn right immediately to 280. Prinair 277. Prinair 277, do you read? Turn to 280. [Instructor Read returned to the approach scope.]

READ: Prinair 277, Prinair 277, San Juan approach. Do you read? Do you read?

The next day the wreckage of Prinair 277 was found 500 feet below the hills at Sierra De Luquillo. All nineteen persons on board perished. Some interesting findings resulting from the accident:

1. A test flight was flown over the identical course as that of ill-fated Flight 277. FAA radar technician Ruben Powell concluded that there was a definite blind spot on the radarscope in the vicinity of the crash.
2. San Juan tower and approach control is authorized for 31 qualified men. They presently have 29, and 4 of these are students.
3. All controllers who handled the flight from takeoff to crash were students.
4. The approach control communications system was in error. Both the approach controller and the coordinator could be heard simultaneously by pilots on the same frequency.
5. Prinair aircraft are not equipped with transponders, thereby making their targets on the scope useless during precipitation.
6. The approach control instructor was carrying the double burden of instructor and coordinator.

George F. Owen, FAA tower chief at San Juan, put out the following notice on April 2, 1969:

> *Subject:* Restrictions to use radar in vectoring aircraft in the San Juan terminal area.
>
> Radar shall not be used below a line shown on the attached local chart. This is deemed necessary due to possible aircraft target fade.

The pressure exerted from airlines is not only on the pilot group, but on all the parameters that may stand in the path of economic gain. All scheduled carriers belong to the Air Transport Association (ATA). They are banded together in a tight group with one goal in mind: to make money. The force exerted on the FAA by their powerful lobby is criminal. Let alone, the FAA might possibly do a decent job. But under the fear of the airlines, they generally do as they are told.

Every minute a small jet is on the ground costs the company at least $5.50. In 1968 delays cost the airlines $80 million, more than twice the cost of delays in 1964. It is easy to see why the airlines, and in turn the controllers and supervisors, are continually forcing traffic to move.

Wake turbulence has been with us for as long as man has flown. For years pilots tried to keep clear of the prop wash of the plane ahead. Many accidents have been caused by flying through the wash of a plane that has just passed—or at least that was the assumption. Engineers discovered that while engine wake did cause trouble, the lion's share was produced by wing vortices. A plane produces two of these baby cyclones, one at each wing tip. Each vortex spirals in a different direction, and their potential is in proportion to the speed and weight of the aircraft. The larger and heavier transports produce the most disturbance and they do it during takeoff and landing. The wake behind a 707 has the velocity of a hurricane; the wake of a 747 is 10 percent higher. The book says that aircraft on final approach must be at least three miles apart. On takeoff this is cut to a few thousand feet.

The treacherous waves of wake have been known to linger behind an aircraft for as long as five minutes and this with a good wind blowing. I will never forget an incident that happened to me on final approach at JFK. I was flying a Boeing 727 and was spaced at three miles behind a DC-8. I was at approach speed when, without a word of a warning, I was suddenly pitched up in a 90-degree bank. There is no question in my mind that the vortex from the plane ahead nearly flipped me on my back. From that day on I became a believer; I always request a five-mile separation.

"Cleared for takeoff, caution wake turbulence" is a familiar phrase used by men in the towers. There is only one caution: wait until traffic has subsided, a virtual impossibility at today's crowded airports.

On December 27, 1968, Captain Marvin A. Payne was in command of North Central's Flight 458, a Convair 580 from Minneapolis to Chicago. As the trip neared Chicago the center handed the flight off to O'Hare approach control. The weather at the time of the approach was barely minimum with a light wind of six knots. First Officer LeValley initiated the call.

LEVALLEY: O'Hare approach, Flight 458 with you.
APPROACH: North Central 458, turn right heading 120, intercept the 14R ILS, fly it inbound, cleared for approach, speed 160 til Romeo, RVR 4,000, position from Romeo is 14 miles.
LEVALLEY: Roger, North Central 458, that heading 120 takeover on approach, 160 to Romeo.
APPROACH: Roger.
APPROACH: RVR [runway visual range or the number of feet that a pilot will expect to see down the runway when he breaks out of the clouds] 14 R North Central 458—2,600 feet.
LEVALLEY: OK.
APPROACH: Flight 458, you are 3 miles behind a 727, that's 4 from the marker, and the tower is 118.1 at Romeo.
LEVALLEY: 118.1 at the marker.
APPROACH: Roger.

The final portion of the check list was completed, and all that remained was to spot the runway visually. After changing to the tower frequency the first officer transmitted.

LEVALLEY: North Central 458 is by the outer marker.
TOWER: North Central 458 number 2 for 14R. The RVR 4,500.
LEVALLEY: OK.
TOWER: North Central 458 cleared to land 14 Right.
LEVALLEY: 458.

That was the last transmission from Flight 458. In less than two minutes 52,000 pounds of metal would slam into a hangar, killing 24 of the 45 on board.

It was found that the plane was precisely on course; the crew functioned in a most professional manner; and the aircraft was performing perfectly. What happened in the last few seconds of the approach has happened before and will most assuredly happen again.

The Air Line Pilots Association reported: "The ALPA committee finds that the probable cause of this accident was insufficient airspace separation between North Central Airlines Flight 458 and the preceding landing Boeing 727. This resulted in an upset of the North Central Air Lines CV 580 when it encountered wingtip vortices generated by the 727 precipitating an unavoidable loss of control at an altitude from which recovery was not possible."

Both ALPA and PATCO recommend an increase of the current three-mile separation standard to five miles. To date this regulation has not been changed, except for spacing behind jets that are considered heavy (over 300,000 pounds).

The working conditions of the air traffic controller have improved somewhat. No longer must he take a bite from a cold sandwich while stuck at his position for hours at a time. The staffing of the nation's twenty-three en route centers has reached a point where they're almost up to strength. The number of on-the-job trainees is not nearly what it was a few years ago. However, the terminals where the tower operators and approach controllers work are still grossly understaffed. O'Hare, the busiest tower in the U.S., is still 40 percent below its quota of controllers.

Let's just visit a few of the terminal facilities around the country and see what problems were encountered in 1974:

DALLAS, TEXAS: *The tower has been operating for eighteen months with half the authorized staff.*

ALLEGHENY COUNTY AIRPORT, PITTSBURGH, PENNSYLVANIA: *Runway lights for runway 23 have been out of service since January, 1973.*

ALBANY, NEW YORK: *A reported blind spot for radio frequency 126.6MHZ exists from 15 to 24 miles northwest of Albany. On March 6, 1974, the FAA flight-checked their surveillance radar and found it unsatisfactory because of inaccurate map alignment.*

AMARILLO, TEXAS: *The tower supervisor allowed a non-qualified trainee to work a position by himself. The full complement of workers calls for twenty-four controllers. They presently have fourteen aboard.*

BALTIMORE, MARYLAND: *Numerous discrepancies in the FAA log concerning weak radar from both channels with video maps jumping out of calibration by as much as two miles.*

BRADLEY, CONNECTICUT: *A faulty switch that controls the intensity of the runway lights has been written up for two years. When turning this switch to step down the intensity of the lights, it is possible for them to go out completely. Maintenance people perform tests on the ILS and radio frequencies as often as four times a week. This is done at peak periods without warning.*

THE COMMON IFR ROOM, NEW YORK, NEW YORK: *This is the control center for all flights into Kennedy, Newark, and La Guardia, plus numerous satellite airports. Although it is one of the busiest facilities in the world, it is understaffed and in order for a man to obtain relief from position for personal reasons and meals, it is necessary for another man to work his positions. Not a safe operation in such a busy facility.*

STAPELTON, DENVER: *Discrepancies in the ARTS 111 radar including radar failure, strobing and inability to affect computer handoffs are numerous here. Rain usually puts the ILS and omni off the air. The typical complement of controllers is nineteen, but at times there are only fifteen men on duty, seven of whom are journeymen, and four are trainees.*

BAER FIELD, INDIANA: *Fourteen journeymen controllers provide services that twenty-four are assigned to. A controller was forced to work two positions and monitor a trainee as well. Men have been forced to sit at busy traffic positions for as long as six hours without relief.*

DULLES, VIRGINIA: *Poor radar coverage and numerous ARTS 111 display problems are common. On May 14, 1974, a complete power failure resulted in radio and radar outages. The back-up power system failed for four minutes, during which time no voice communications were possible. This problem has been reported several times in the last few years. Staffing is also inadequate.*

HOUSTON, TEXAS: *ARTS 111 radar has been reported as unreliable most of the time for the past two years.*

LONG BEACH, CALIFORNIA: *A Boeing 727 was on final approach when the pilot spotted a man calmly riding his bicycle down the active runway.*

RENO, NEVADA: *The radar became partially inoperative at a time when the surrounding 10,000-foot mountains were obscured by clouds. The problem was not rectified for two days because of a maintenance policy that prevents repairs for only partial failures.*

STANDIFORD, KENTUCKY: *All electric power was lost while forty aircraft were being radar-vectored. Radar and radios were off the air. The back-up battery-operated radio failed after two transmissions. Prior to the failure three aircraft had declared an emergency. This failure lasted for 13 minutes. Fortunately the weather was good.*

SEATTLE, WASHINGTON: *Staffing calls for forty men. They have only thirty-three.*

SAN JUAN, PUERTO RICO: *There are only twenty controllers to man the positions of an authorized thirty-two. In June the departure control radar was shut down for several days due to a shortage of controllers.*

As you can see the system still has a long way to go. The FAA budget for October 1975 has no money allotted to hire and train the much needed controllers. A PATCO controller from the Des Moines tower summed up the situation nicely when he said, "This place needs controllers, not an explanation of why we don't have any. Whose neck is hanging out when two airplanes share the same airspace? Two dead pilots and an unemployed controller."

Chapter Fourteen

THE KILLER ALTIMETER

YEARS ago when flight instruments were first introduced they were regarded by the pilots as just another gauge in the cockpit. Pilots kept from falling out of the sky simply by listening to the wind as it whistled through the struts. They "flew by the seat of their pants." If their body was displaced in a certain direction they were either skidding or slipping. With the advent of weather flying, faith in your instruments became a necessity.

Before getting into the costly tricks that some instruments have played on pilots, I would like to touch on the systems that make them function. Nearly all the systems have changed very little since World War II. In fact, the static system got its start in World War I. This same system, with only minor changes, is what keeps our modern jets from crashing—well, almost.

I am sure you can recall putting your arm out the car window when you were a child and having the force of the wind push it backward. The difference in air pressure from the stationary air in the car to the rush of air outside produced the movement. This is the principle that operates the airspeed indicator.

Extending forward into the airstream on all planes is a small diameter hollow tube. The other end is fastened to a small bellows in the cockpit instrument (similar to a barometer used in the home). The difference in pressure between the air rushing into one tube (pitot pressure) and another tube that is vented at the side of the aircraft (static pressure) is what makes the airspeed and mach indicator function. This instrument tells the pilot his speed through the air—a vital necessity if he is to avoid exceeding the maximum structural speed, or getting into a deadly stall while landing.

While the airspeed indicator uses two pressures to function, the altimeter and vertical speed gauges (velocity up or down measured in feet per minute) use only the static line. A tiny metal bellows can be found in the heart of all three instruments. Their outside cases are all sealed airtight.

As the aircraft climbs, the atmosphere pressure decreases. The pressure on the outside of the bellows grows less and it expands; the opposite is true when a descent is made. A series of gears and levers transmit this action to pointers on the face of the dial. For the needles to revolve around the case only a small movement of the bellows is needed. From ground level to 80,000 feet requires an expansion of less than one-quarter inch. If you changed your house barometer to read in feet instead of rain, cloudy, etc., you would have a parallel to the airplane altimeter.

The small opening at the ends of both air tubes are subjected to all the elements. A means of applying heat to both tubes should be necessary. The pipe that faces out in front of the aircraft and supplies information to the airspeed indicator is heated. The static ports are not. They are exposed to all the rain, snow, and ice the plane encounters. When this tiny line is obstructed, the readout of all three instruments will be erroneous.

At one time both static ports were heated, especially on Air

Force planes. Today only a very few commercial jets have both lines heated. The Boeing Company has produced many fine transports but only two models, the 727 and 737, have heated static ports. The static ports are supposedly placed in a location that is relatively ice free. (On occasion ice has been known to form on an airplane where engineers predict that it cannot.)

An aerodynamic staff specialist from the McDonnell Douglas Corporation was testifying before the hearing board concerning the crash of a DC-9. He reviewed the results of tests on the static system and the effects of water running over them. He stated that complete blockage of the static system would induce errors of 100 feet and 7 knots on the airspeed. An FAA systems and equipment engineer also testifying at the hearing made the statement that there are no standards to determine the design criteria of static systems. You can see how concerned the FAA is about one of the most vital systems in the aircraft.

I have on occasion seen as much as an 800-foot differential between my altimeter and the reading on that of my co-pilot. There is no way to check which one is right in the air—and when you get on the ground it may be too late.

On November 20, 1967, a TWA four-engine jet crashed a mile short of its destination. Over fifty people died. The pilot was making a nonprecision approach. It was the only one available at the time of the accident, but without a precision approach it was impossible to check the altimeter. Low-hanging clouds dropped rain and sleet, a good mixture to block off the life-giving air necessary to keep the altimeters functioning properly.

The ALPA accident report states that the cause of the disaster was, in fact, just that. They recommended that all static ports on all jet aircraft be heated, to prevent ice formation. Also suggested was that static systems be designed to self-drain, immediately, of any water accumulation encountered in flight.

Static-system checks vary from one airline to another. They certainly are not frequent enough. If an aircraft remains overnight at a good-size maintenance base, the static ports may be checked, but a thorough examination of the system won't be performed for months. Federal Regulation 91.170 is probably as lax as it can be. It states that any airplane operating under instrument conditions must have the altimeters and related lines checked every two years.

The pitot systems on all Air Force aircraft are covered as soon as the mission is terminated. This keeps dirt and bugs from entering the lines. One jet pilot was killed because he took off and his airspeed indicator failed. A hornet was discovered lodged in the tubes. Since then it has been mandatory to protect the system. Air carriers do not take advantage of this safety precaution.

Airline and airplane manufacturers recently conducted a series of tests on static ports. They experimented with lines of greater diameter than the ones in present use. It was proved that if the lines were increased in size from .047 inches to .125 inches the possibility of water forming a capillary barrier over the ports would be eliminated. Also, this larger orifice would permit nearly instantaneous draining of water in the lines. As of this date, few modifications have been made on any of the existing systems.

There are numerous altimeters on the market today, some good, some bad. We will be discussing two of the most dangerous models: first, the altimeter with three pointers.

A good parallel when learning to read the altimeter would be to compare it to a clock with a sweep-second hand. Both the clock and the altimeter have a common scale. The watch is divided into twelve equal parts, the altimeter into ten. The second hand on the clock represents the 10,000-foot pointer. The small hand is equivalent to the 1,000-foot pointer, and the large hand shows hundreds of feet. It takes one complete revolution of the 100-foot pointer before the 1,000-foot pointer moves to one. One complete revolution of this pointer and the third needle points to 10,000 feet. When the three pointers line up in various positions the altimeter must be *studied* rather than merely read. Even then it is possible to read it incorrectly.

The pilot of a 707 was cleared for an approach at an overseas terminal. Both altimeters were set and functioning correctly. The skipper was flying in the clouds and relying solely on his instruments. Each time he got down to his minimum approach altitude of 1,000 feet, he saw nothing but clouds—in spite of the fact that the tower advised that the ceiling was 5,000 feet, well above his minimum altitude. He made another attempt, still no runway in sight. On the third try he noticed the smallest hand on the altimeter just before it was obscured by one of the other pointers. He had been making a 10,000-foot error in reading his altimeter. Each time he thought he was at 1,000 feet he was in reality at 10,000. This is a trait common to the three-handed instrument.

On the evening of August 16, 1965, Captain Melvill W. Towle, forty-three, was in command of United's Flight 389, a Boeing 727 en route from New York's LaGuardia to Chicago's O'Hare. Captain Towle received descent clearance from ATC and left his assigned altitude of 35,000 feet. At 9:21 P.M. witnesses saw the plane come through the overcast and dive into Lake Michigan. All thirty on board were killed instantly.

No evidence to date substantiates any preimpact difficulties with the aircraft. The probable cause has not been found. The NTSB related that the two altimeters on this craft were the three-pointer type. They stated, "The three-pointer altimeter is the most susceptible to misreading of any of the four types presently in use in commercial aviation."

It is the duty of the NTSB to recommend to the FAA any changes in regulations, procedures, or equipment that they consider dangerous. The outlawing of a killer altimeter should have been made years ago. Many of the airlines have gone to the drum-type altimeter, but there are still carriers with all-jet fleets that are 100 percent equipped with the old three-pointers. The FAA remains silent on this issue.

Years ago, when the Lockheed Electra was introduced, a recently-checked-out captain was making a nonprecision approach into LaGuardia when he crashed. He was using one of the latest drum-type altimeters but he had had little formal training on reading the new instrument. The accident board partly attributed the crash to this lack of training. The FAA still has no formal requirements to make certain that a pilot fully understands a new device. Because of the newly required altitude-alerting system, many airlines have changed to the drum altimeters. (There was a mandatory ruling this year that required all airliners to have an altitude-alerting device installed, a light and buzzer that will activate, warning a pilot that he is approaching his assigned altitude. There is no TSO [technical standard order, or seal of approval by the FAA] concerning this alerting device. It could be made and installed by the village fix-it shop.) And there are pilots flying today who "learn as they go." By that I mean that the FAA allows a crew to fly with an altimeter they have seen before only as a picture in a book. Most of the instruments and components in an airliner are certified by the FAA, but not all.

The only way a pilot can check his altimeter on the ground is by getting a setting from the tower. As the altimeter is nothing more than a weather barometer, it stands to reason that any changes in the weather will affect it. Each instrument has a small numbered window on its face. By turning a knob, sea-level pressure can be set. This knob when rotated will move the position of the pointers on the face of the gauge. If you were to sit in the cockpit of a parked aircraft and watch the altimeter as a storm passes, you would notice the 100-foot pointer slowly indicate a climb of 300 or 400 feet, depending on the severity of the storm. As the storm passes the pressure increases and the needles start a slow descent. The big danger of flying into a low-pressure or low-temperature area is that the altimeter reads higher than you actually are.

When he is ready to begin a flight the pilot always calls the tower for an altimeter setting. Once set, the gauge can be checked against the field elevation. For instance, field elevation at O'Hare is 667 feet. When set correctly, that is exactly where the altimeter should be pointing. It is possible to start a flight with an altimeter error of 10,000 feet. Should the departure airport's elevation above sea level be such that the 10,000-foot pointer is completely covered, you are set for trouble.

If for some reason maintenance was turning the setting dial, it is possible to turn it until the 10,000 foot needle is pointing to 1 (or 10,000 feet). Here the setting scale returns to normal. When the pilot calls for his altimeter check from the tower it is conceivable for the 10,000-foot pointer to be hidden by one of the other hands. If this error is not detected prior to takeoff, it can go unnoticed until the accident team arrives at the scene of the wreckage.

In the United States altimeter settings are usually given with the first digit missing. For instance, if the proper setting is 29.92 (inches of mercury) the first digit is omitted, making it 992. This is done strictly for brevity on the radio, and all U.S. pilots are cognizant of this. In foreign countries the altimeter setting is given in millibars. The problem here is twofold. American pilots who fly out of the country might interpret the 992 setting to mean inches, when in reality it refers to millibars; this could make the pilot dangerously low. The opposite would be true if a foreign pilot came to the United States. Even though a pilot knows better, habit can become a killer.

Another altimeter fault is that the pointers occasionally stick. On the older prop jets there was no problem concerning necessary vibration. The jet aircraft is relatively vibration free. A means was devised to provide a steady tapping on the instrument case, or the whole instrument panel. All Air Force jets have this feature on their panels. Some airlines have electric instrument thumpers, and do not fly unless at least one of them is functioning. Other airlines have no such device at all.

In October 1966, Captain Charles C. Warren, forty-two, and seventeen others, were killed instantly when the DC-9 he was piloting rammed a rain-soaked mountain. A test airplane, similar to the one that crashed, was immediately dispatched to re-enact the path of the first. It was still raining when the check ship made its approaches. They recorded three instances of "sticky" altimeter. The needle on one gauge moved in erratic jumps of between 50 and 70 feet. The biggest discrepancy was observed at the minimum altitude. Here one altimeter lagged behind as much as 100 feet, and then tried to catch up in 40- to 60-foot jumps. The reason for this accident remains unsolved; all that is known is that the pilot descended too low. There is still no regulation making an altimeter vibrator mandatory.

In recent years a new and better type of altimeter has been developed, the single-pointer drum-type altimeter. This eliminated two of the pointers that were necessary on the older model. Now a pilot can find out his altitude from two rather than three indicators. Reading this altimeter is easier than reading the old one, but it is still deficient; instead of a 10,000-foot error, the discrepancy has been reduced to 1,000.

This altimeter operates from the same ancient static system as its predecessor. The main change is in the instrument readout. The single pointer registers in hundreds of feet. The 1,000-foot information is read directly from a tape that is visible through a small window. This tape is numbered from 1 to 80. As the height of the craft is changed, this drum slowly rotates and displays three consecutive numbers. An index on the face is the reference mark of these numbers.

As the 100-foot single pointer rotates completely around the scale, the drum turns to the next number, indicating the height in thousands of feet. It is imperative when scanning this gauge that only the number directly below the index be used. For example, when the 100-foot pointer is showing 900 feet, the drum index shows one number slightly above the index mark and the zero below it. A misinterpretation is highly feasible at this point. A quick glance from the pilot, and he would swear that he is at 1,900 feet rather than his actual 900 feet.

Captain William J. O'Neill, thirty-nine, had flown for American Airlines since 1951. He was upgraded to captain in April 1957. His logbook total showed 14,400 hours. Having passed all the necessary tests for a type rating on the 727, he was now in the final stages of being route qualified. With him on Flight 383 was Captain David J. Teelin, his instructor, forty-six, occupying the right seat and performing the duties of first officer. Flight 383 was routine until it neared its destination. As they approached Cincinnati the forecast thunderstorms became a reality.

At 8:55 the check pilot radioed Cincinnati approach control: "Out of 5 for 4 and how about a control VFR? We have the airport."

The approach controller replied: "Continue to the airport and cleared for a visual approach to runway 18."

At this time the workload in the cockpit was at its peak, especially when attention had to be diverted to keeping the field in sight as well.

The tower called again: "American 383, approach lights, flashes, and runway lights are all high intensity."

"OK," the co-pilot replied as he tried to keep one eye on the field. That was the last radio transmission received. The lives of fifty-eight people were snuffed out two miles short of the airport.

There was no reported malfunction throughout the flight, nor during the approach. The aircraft was equipped with three drum-type altimeters. The NTSB labeled this tragedy "failure of the crew to properly monitor the altimeter during a visual approach."

The air force has banned the drum-type altimeter and do not have it on their latest planes.

Since February 28, 1972, all turbojet aircraft (propeller aircraft experience the same mishaps, but they evidently don't count) have been required to have an approved altitude-alerting system. The type of device is left to the discretion of the various airlines. On this system, either a buzzer or a light will automatically actuate after a flight passes a predetermined altitude. An FAA advisory circular had this to say concerning the warning system: "Altitude-alerting devices may utilize information from the basic altimeter system."

As there are no warning flags on the airspeed, vertical speed, or altimeter to warn the pilot of a malfunction, any device that depends on potential erroneous static information cannot be relied on. The best safety device for vertical clearance close to the ground is the radar altimeter. It functions in the same fashion as all radar except that its beam is directed toward the ground. At a predetermined altitude, a light flashes a warning to the pilot.

The 747 is equipped with one of the latest and best altimeters ever produced. It contains one pointer that registers altitude from 0 to 1,000 feet. From this level up to its maximum indication of 50,000 feet, it has a digital readout. No longer is it necessary to examine the pointer on its face; it can be seen easily. As altitude is changed it can be read directly from a window similar to the one found on an automobile to show the total mileage. It takes most of the guesswork out of altimetry. When this instrument is used in conjunction with a radar altimeter and an altitude device, it becomes nearly foolproof. Some airlines already have these units installed on their planes. Others wouldn't put two pilots in the cockpit if it were not mandatory.

The NTSB is cognizant of the altimeter problem. In their annual report to Congress they state: "The reassessment of altimetry systems with particular regard to their susceptibility to insidious interference by forms of precipitation needs to be the subject of attention by the highest level of aeronautical research facilities and personnel." They still have not recommended to the FAA that the radar altimeter should be utilized as a backup on all airliners.

Hundreds of people have been killed for lack of safety devices that are available today. How much longer will it go on?

The cockpit of a Boeing 747 jetliner. The all-important "little black box," the flight data recorder and its components, often the only clues to an aviation disaster.

The human and aircraft remains of the Turkish Airlines DC10 which crashed in a forest near Paris on March 3, 1974. All 346 passengers and crew were killed in what is still listed as the worst disaster in civil aviation history. TWA flight 514 from Indianapolis to Washington, D.C., ended in tragedy on December 1, 1974, when the Boeing 727 crashed in bad weather near Dulles airport killing all 93 persons aboard.

This New York mid-air collision on December 16, 1960, resulted in the deaths of 128 passengers. A TWA Super Constellation collided with a United Airlines DC-8. The planes were attempting to make instrumental landings at the La Guardia and JFK airports, respectively, when tragedy struck. The TWA plane fell in a field in Staten Island, the United jet landed in a residential street in Brooklyn.

Four tragic crashes . . . 227 people killed. A sheriff's rescue team probes wreckage of an Air West jetliner lying in a mountain gorge near Los Angeles on June 6, 1971. The plane had collided with a Marine Corps Phantom F4 jet fighter, carrying 49 people to their deaths. A United Airlines DC-8 jet lies a blackened $5-million ruin after crashing into Denver's Stapleton Field. Seventeen died in the flaming wreckage. A Pan American Boeing 707 jetliner crashed into a Samoan jungle on January 31, 1974. Ninety-two occupants of the jet were killed. An Eastern DC-9 crashed during a routine landing in Charlotte, N.C., on September 9, 1974. Fortunately, these were among the 13 people who survived the fiery crash, 69 others died.

A rural area near the Bradford, Pennsylvania, airport, was the scene of two fatal crashes, both within weeks of each other. The first, on Christmas Eve, 1968, killed 20 people when an Allegheny Airlines Convair 580 crashed. On January 6, 1969, another Allegheny Convair 580 crashed nearby while attempting a landing. Eleven people were killed. On December 4, 1971, a small private aircraft collided with a huge Eastern DC-9 jet over Raleigh, N.C. The jetliner landed safely with minor damage to its fuselage. The light plane crashed, killing both occupants.

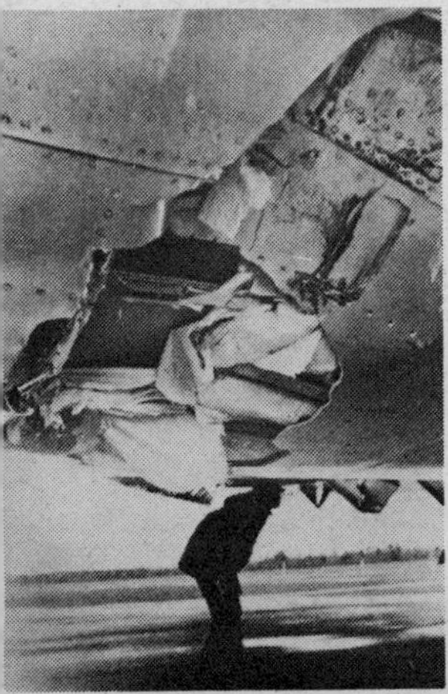

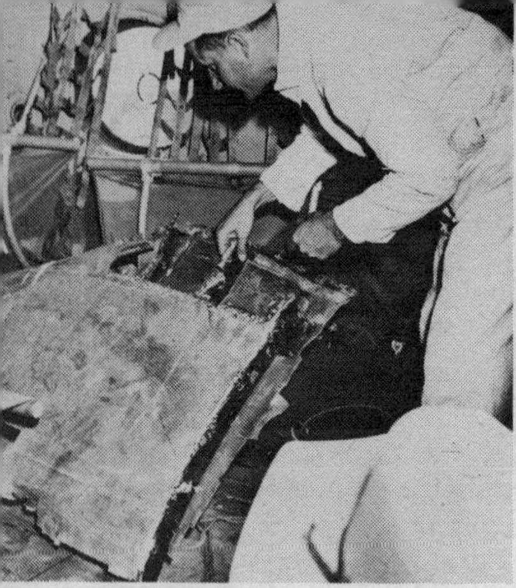

The National Transportation Safety Board, the Federal Aviation Administration, The Air Line Pilots Association, the Professional Air Traffic Controllers Organization, and other governmental and corporate investigators study every fatal crash to determine causes and, hopefully, discover preventive measures. Here we see an investigator probing a piece of wreckage from the TWA Super-Constellation that collided with a United DC-7 over the Grand Canyon on June 30, 1956, 128 people were killed. In a hangar at New York's JFK airport examiners attempt to reconstruct the wreckage of an Eastern Constellation that collided with a TWA jetliner near New York City on December 4, 1965. Officials probe wreckage area in Henderson, N.C., where a Piedmont Airlines 727 jet collided with a small private craft. The crash took 82 lives, left no survivors.

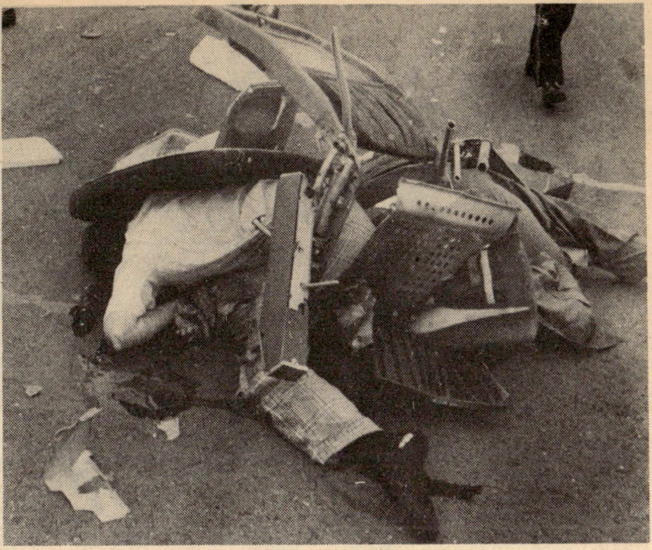

The grim scene at Boston's Logan Airport on July 31, 1973, when a Delta Airlines DC-9 crashed while landing in a heavy fog. The huge craft disintegrated upon impact, killing 88 people.

Air traffic controllers at work guiding airliners in and out of our busy airports. The author as "Captain X," when *Safety Last* was first published, and later, when he revealed himself as Captain Brian Power-Waters.

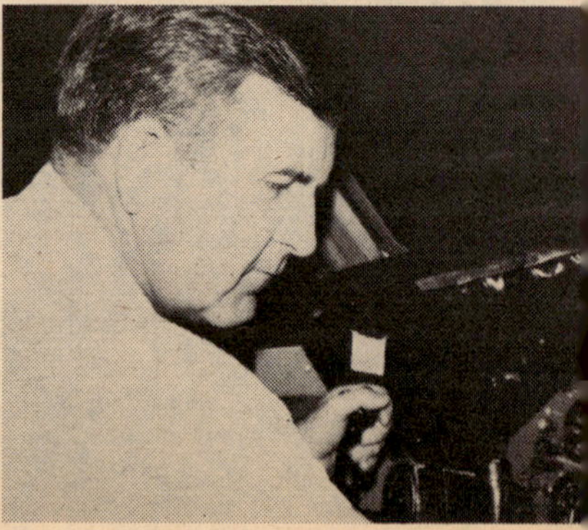

Chapter Fifteen

SEEING IS NOT BELIEVING

TWA'S CAPTAIN IVAN SPONG is a "killer." That's what the CAB investigating team said. They stated that Spong intentionally flew his aircraft into Sandia Mountain near Albuquerque, New Mexico, and took the lives of fifteen other people with him. The "killer's" widow was inundated with malicious phone calls. People were only too willing to take the word of the investigating team and vent their animosity on a pilot's widow. Even the Spongs' son was so harassed by his schoolmates that he developed an ulcer.

Both Mrs. Spong and Captain Larry DeCelles, another TWA pilot and a family friend, were convinced that Ivan Spong was a victim of circumstances. Captain DeCelles had warned TWA *before* the accident that the flux-gate compass, a more reliable direction indicator than the magnetic compass, could at times be in error.

TWA and Captain DeCelles worked diligently for years to clear Spong's name. Captain DeCelles, aided by his fellow pilots, collected many cases of flux-gate malfunction, but the stubborn board would do no more than strike out the word "intentional" from its accident report. Five years after the crash in 1960, there was an overwhelming mass of conclusive evidence concerning erroneous instrument readings. The CAB backed down (a rare move) and cleared the pilot's name completely.

The compass mounted on the dashboard of many cars is essentially the same in looks, size, and operation as the compass in a modern jet. The next time you drive, notice the erroneous readings your auto compass shows whenever you turn a corner or speed up. All these errors are prevalent in a modern magnetic compass. In fact, on a plane they are more accentuated, and in rough air the compass oscillates to such a degree it cannot be relied on.

A means for producing a guidance system that would not be susceptible to these errors became a necessity. The first instrument that was widely used for direction was called the "directional gyro." This instrument had no magnetic capabilities in itself; it could be set to read a given direction from the magnetic compass. Because of the great speed of its inner gyro, it would hold a specific course without drifting off for as long as fifteen minutes. It was necessary to check this gyro reading with that of the magnetic compass continually, thereby increasing the workload in the cockpit.

The "flux gate" was a big improvement over the straight gyro. This system was more sophisticated and cut down on the workload. Two master units were placed inside the wings, near the tips. Here the magnetic disturbance would be minimal. These indicators were the brains of the system. Through a minute electric current they transmitted their direction to an amplifier which boosted the signal to a compass card in the cockpit. This system, when it was operating normally, was fairly accurate. However, when it malfunctioned there was no warning given to the pilot. The only way a captain could check on this instrument was to compare his indicator with the first officer's gauge, or with the old-fashioned wet compass.

On April 4, 1964, Raymond Hourihan was hired as a co-pilot by Mohawk Airlines. It was a position he had waited for most of his life. He studied diligently and was one of the most capable pilots on the line. In 1968, at thirty-one, he successfully completed all phases of his captain's upgrading and was issued his ATR (Air Transport Rating).

Captain Hourihan was assigned as a standby pilot at LaGuardia. Here he flew for six months as a captain on the turboprop Fairchilds. As luck would have it, LaGuardia base received a cut in scheduling and Hourihan returned to his main home in New Hampshire. Seniority would not allow him to fly as a captain out of Boston. He reverted back to co-pilot on the BAC-111 and was assigned a regular run in that capacity.

The FAA allows pilots to switch from one type of aircraft to another—as long as they are qualified on each type, they're legal. Being legal and being safe are two entirely different matters.

On November 19, 1969, Captain Hourihan received a call from his airline's crew scheduler offering him a captain's trip on the Fairchild, which he accepted. When he reported to operations at LaGuardia he met for the first time his co-pilot John P. Morrow, thirty-one. Six months earlier John was flying fighters in Vietnam. Although his logbook totaled nearly 4,000 hours, airline flying was relatively new to him.

Flight 411 left on schedule and touched down at Albany, New York, an interim stop. The turbulence encountered over the last portion of the trip prompted Captain Hourihan to deplane at Albany and recheck the weather en route to Glens Falls. At that time there were no less than two aircraft grounded at Albany for severe turbulence checks. Both the pilots of these planes came in from the west and experienced a thrashing about they will never forget. They both stopped their flights to have maintenance give their crafts a thorough check. It is not known if either of these pilots warned Ray of the impending danger approaching Albany. Captain Hourihan was known as a safety-conscious flyer, and had he been told of the approaching danger he most likely would have canceled his flight.

Flight 411 taxied out and F/O Morrow copied their airways

routing from the tower: "Mohawk 411 cleared for an approach to Glens Falls airport. Cruise 3,000 feet, left turn after takeoff to on course contact departure on 126.6 squawk 1,000 low."

After the readback, the before-takeoff check was accomplished and through rainstreaked windshield the green runway threshold lights could be seen. Takeoff clearance received, Flight 411 taxied onto the active runway. The engines were advanced to maximum power, and as the turbines reached a screaming pitch the brakes were released and 411 was on her way.

Glens Falls airport has no radar, no approach lights, no ILS, no tower, no runway overrun for its minimal 5,000-foot strip, and on this occasion a straight-in nonprecision omni approach would not be available. When the wind is light or from the north a straight-in approach can be used. This night, because of the strong winds from the south, a *circling* approach would be mandatory.

In aviation there is probably only one thing more dangerous than shooting a circling approach: doing it on one engine. During this letdown a pilot is allowed to descend to circling limits, in this case 500 feet above the ground with a visibility of one mile. As soon as the runway is visible an immediate turn is made away from it so that appropriate spacing can be made to enable a landing at the other end. During this horrendous period the pilot has no outside guidance, only the runway lights which are barely visible between patches of clouds. He must keep changing focus from long range (the ground) to short range (his instruments). The attitude, airspeed, altitude, check list, and runway are just a few of the lifesaving items that must be dealt with simultaneously and accurately.

The distance from Albany to Glens Falls is just over 40 miles. North of the field, jutting out of Lake George, sits Pilot Knob Mountain—this night shrouded by rain and clouds.

"John, tell Albany we're going to change over to Glens Falls radio."

"OK, Ray. Boy, it's as rough as a cob."

"You can say that again, hope this thing stays in one piece."

The rain was forming a solid sheet of water over the windscreen, and the airspeed and altimeter needles were doing a dance of their own.

"John, tell them we're starting down. Hey! Look at the compass. Mine shows 080 and we're right on the 186 radical."

After a careful check of the vibrating instrument panels First Officer Morrow said, "Mine reads 330."

"We've got real trouble now, as if this turbulence weren't bad enough."

The squall line that hit Ray's plane was now pushing it along with an additional 80-knot tailwind. Knob Mountain was getting nearer every second.

"Call Albany radar. Tell them we want a steer and . . ."

CRASH!

At 8:20 P.M. Flight 411 rammed into a solid rock cliff and exploded. Fourteen more lives wasted. The aircraft was almost completely destroyed by fire, along with it the telltale clues of the cockpit voice recorder. What you have just read is my reconstruction of the crash. The NTSB labeled the case "pilot error."

If Glens Falls had been radar equipped, this accident might never have happened, and if there had been warning devices on the direction indicators, their faulty indications could have been detected earlier. Two very important *ifs* that still have not been rectified.

The direction system on a modern jet is basically the same as it was years ago; the components have more sophisticated names now, but the faults remain. The attitude and direction indicators nearly all require both AC and DC for their operation. At least these instruments do have a small red flag that extends into view when a part of the system is not functioning. On the surface that sounds like a foolproof system. Nothing could be further from the truth. These warning flags are held out of sight by electricity. When the power is lowered to a specific level, out pops the flag. Here's the rub: as long as the power is continuous or only momentarily interrupted, there will be no warning, even if the instrument is 180 degrees out of phase.

The pilot of a DC-8 was executing an ILS approach at a metropolitan airport. Upon reaching his minimum altitude a missed approach was initiated. During the climbout the tower advised a turn to 090 degrees, maintain 2,000 feet, and contact departure control. This was acknowledged and complied with.

Departure control was channeled in and called: "Departure, 318 is with you 090 degrees 2,000."

"Roger, squawk ident . . . I show you holding 180 degrees. Check your heading."

The crew had experienced a failure that "*couldn't happen*."

The captain and first officer's instruments both have separate power sources. This supposedly foolproof system should enable one instrument to be checked against the other. Whenever the power to either instrument is interrupted, even for a fraction of a second, the position of the compass is immediately affected and the master unit in the wing knows that its lifegiving current was shut off and it is showing the wrong heading. In order to rectify this error it turns and searches for the correct direction.

In the episode just mentioned, the electric power for the whole aircraft was momentarily curtailed. Whenever this happens in any position other than straight-and-level flight, the compass goes wild. By losing power to both instruments at the same time they both fast slaved and stopped on the same heading—leaving only the ancient liquid compass as a check. Any pilot who has four compass indicators all lined up together has no reason to check his standby compass.

Imagine the potential disaster that a system like this could cause, especially if there is a mountain out there waiting for you. If it were not for the radar controller a very dangerous situation could have developed. Even in a relatively flat area there is still the chance of running into another aircraft.

If one instrument were to be singled out as more important than any other, it would be the *flight director*. The early attitude indicators were run by vacuum-driven gyroscopes similar in principle to the small gyroscopes children play with. A rapidly spinning gyro maintains its relative position in space, thereby giving the pilot a steady reference. Nowadays the old horizon has a new name and power source. It's called a flight director, and like the compass, it derives its power from two sources requiring both AC and DC current.

A small, flat, triangular-shaped piece of metal in the center of the instrument represents the relative position of the airplane. A tape behind this piece depicts the horizon. There are usually four lines running horizontally across this tape: the horizon splits the middle line and above this are three shorter lines representing pitch up of 5, 10, and 15 degrees. When you are flying straight and level, the top point of the triangle is usually splitting the horizon mark. Every movement the aircraft makes—up, down, left or right—is displayed on this tape.

When you are making an ILS, an added attraction becomes available. By turning a knob to the approach mode two V-shaped bars (command bars) descend from the top of the case and give you three-directional readout to capture and follow the localizer and glide path to the runway. All that remains is to keep the triangle inside and flush against these bars; it should guide you to the runway.

Coupled approaches are also possible with this system. A pilot can select autopilot mode and "George" will fly the approach automatically. Some of these instruments, however, have been known to guide an aircraft in what seems like a normal descent—right into the ground or water.

Let's take a look at just how well these systems perform. A twin-engine jet transport crashes and burns; all on board are killed. In the ensuing investigation it is discovered that the plane experienced an electric failure during a night flight through turbulence; the power failure, with a resultant loss of both primary attitude and heading indicators, caused an uncontrollable condition.

The NTSB recommended, "The Board believes that various modifications which have been accomplished in the instrumentation and electrical system of the Lear jet since this accident negates the requirement for additional recommendations at this time. Among the improvements was the installation of an attitude indicator powered by a source separate from the aircraft primary electric system."

Six months after this accident another Lear jet crashed under exactly the same circumstances. The NTSB issued the same recommendations. This time they took the trouble to send questionnaires to the operators of this aircraft. Fifty percent of the owners claimed they had experienced horizon instrument failures. After the first accident the board recommended that an additional attitude instrument be included in this plane; obviously that recommendation was disregarded in both cases. The FAA could, and should, have made compliance mandatory.

On January 18, 1969, United's Flight 266 dove into the Santa Monica Bay, killing thirty-eight people. The aircraft had just departed Los Angeles and at 800 feet entered a solid overcast. Prior to this trip the plane had operated for three days and a total of 42 flight hours with one of its generators inoperative. Because of heavy flight scheduling, this plane had operated through 28 airports, 23 of which possessed line maintenance. Yet the generator was not repaired. This is perfectly legal according to the ATA's minimum equipment list. (As of this date the list still allows a two-generator operation.)

Two minutes after lift-off United's Flight 266 experienced a fire in the left engine, and it was shut down. The remaining electrical load was reduced, but not in time. The only generator left tripped off the line, and all lights and instruments failed. A standby power switch, when activated, supplies power to the essential instruments. Evidently it failed.

The craft could easily have made a safe landing, however; the power from two engines is sufficient. The problem in this instance was the inability to maintain attitude controls while in the clouds with no instrument references. If the aircraft had been equipped with an automatic electric changeover or an additional horizon with its own power supply, this accident might never have happened.

I have had two horizon failures in the last two years. Luckily both of the malfunctions occurred in good weather, and during light cockpit workloads. We were cruising at our assigned altitude of 31,000 feet. The weather was clear and it was my turn to work the radios while my F/O flew. I had just written a discrepancy in the logbook and put it in its place when I glanced at my flight director and did a double take. It was telling me that we were in a screaming spiral and heading down at a steep angle. A look out the window satisfied my curiosity, and it was confirmed by a check of my co-pilot's gauges. Given different conditions a serious error such as this could go undetected until it was too late.

This system contains warning flags, but once again they only warn of power loss. As long as electricity is available to the instrument, it may be telling you that you are in a climb when in reality your craft may be descending, or that you are in a right turn when in reality your plane is turning in the opposite direction. Emergency situations usually pyramid, and there often isn't sufficient time for more than a quick cross-check.

Here's another classic example of horizon failure. A DC-8 took off into an unstable air mass. The F/O was flying and as he rotated on takeoff the captain's artificial horizon indicated a full 90 degrees nose-up position. They were now in the clouds with only one horizon.

The autopilot was engaged to cut the workload, and to allow more time to concentrate on the dilemma. Now the captain's direction indicator was giving false readings. When he tried to switch his instruments' power source to that of his co-pilot, the switching system proved to be inoperative.

The F/O horizon showed a 20 degree nose-down and a 30 degree right turn. Luckily radar came to the rescue and a safe landing was made. As the engines were reversed on rollout the power supply for the captain's instruments slid out of its rack and the problem became obvious. The wiring for the captain's switching unit along with horizon compass were all contained in the one black box. With this unit damaged, the only instruments left were those of the co-pilot.

The FAA issued a proposed rule-making docket on June 17, 1969, wherein they finally came to the conclusion, after eight years of accidents caused by faulty attitude instruments, that it was time for mandatory backup systems:

> In 1969, an air carrier airplane was involved in an accident and it is known that the airplane experienced a total electrical failure. . . . A warning flag is expected when the instrument malfunctions, but there are many cases of reported discrepancies between the pilots and co-pilots attitude instruments when no warning flags appeared.

The FAA predicted that there would be an average of eight dual instrument failures in the year 1972. Because of the many accidents and the admitted potential of more to come, they finally made a good regulation.

An additional gyro capable of operating for a period of thirty minutes from a separate source of power had to be installed on all turbojet aircraft (props still don't count) by February 8, 1972. The FAA finally made a good rule.

Chapter Sixteen

FARces

A STOP SIGN is seldom put up at a busy intersection until after the accident has happened. It is a terrible thing to realize that the rules governing flight are also fashioned in this way. The public doesn't have a chance.

Sections 61, 91, and 121 are the main Federal Air Regulations (FAR) that scheduled airline pilots must adhere to. Let's discuss them and you can judge for yourself.

Most of the regulations in use today are as ancient and as useless as the landing aids that were put into operation thirty years ago. The 747s operate under the same regulations originally drafted for the DC-3. Over thirty years ago! What follows are a few of the FARces:

61.35 (C). An applicant for an instrument rating must have at least 40 hours of instrument time under actual or *simulated conditions. The time must include at least 20 hours in an airplane.*

The only time requirement for flying a plane solely by the use of instruments is a scant 20 hours. And this time can be simulated: the windshield directly in front of the student is purposely blocked so he cannot see out of the cockpit; his instructor watches other traffic and grades him on his actions. Under this regulation a student pilot gets a license that qualifies him to fly in the worst weather, with only 20 hours' experience at flying blind and possibly no experience in cloud or fog.

121.437 (B). Every co-pilot hired by an airline must have at least a commercial license and an instrument rating. The FAA issues various ratings. If you qualify on an airplane with one engine, you receive a license to fly single-engine only. If you are tested on a multi-engine plane, your license will read multi-engine qualified. As there are no single-engine airlines, why is it that a new man can hold down a job on a multi-engine airline while his license still reads single-engine?

61.47 (C). No airline pilot can fly a trip unless within the preceding 90 days he has made at least three takeoffs and three landings in the type of aircraft to which he is normally assigned. When a pilot goes on two weeks' vacation and comes back for his first trip, it is not uncommon for him to "bounce" the first few landings until he gets the feel of it again. Aviation is a profession that must be practiced frequently. Even the little quirks peculiar to each type of aircraft can be forgotten in a short period. Also, this regulation makes no mention of a required landing at night. What is done easily by day can be tricky at night. Add a low-visibility approach with icy runways and you have an accident in the making.

After a three-month layoff it is conceivable that a pilot may have forgotten the correct start procedure, let alone landing techniques. Regulations that govern Federal Aviation Administration pilots state that they must have at least five landings and takeoffs in 90 days. That is two more than required for an airline pilot and FAA pilots seldom carry passengers. If they do, they must make five takeoffs and landings at night. An airline pilot is not required to make night landings.

On January 18, 1969 a United Air Lines 727 crashed, killing thirty-eight people. The captain had not flown a 727 for forty-seven days, and this was his first flight after being checked ut. Prior to this trip the captain was assigned to a DC-8, a craft hat has little in common with the 727. No regulation exists which limits the different types of airplanes a man can fly in the ame day.

61.47 (C). An airline pilot must get a minimum of two ours of instrument weather flying every six months. Lucky for the traveling public, pilots average about 15 weather ours per month. That is not the case at hand, however. It is onceivable that pilots who fly in the Caribbean area or in a imilar fair weather environment could be caught by this regu-ation. Two hours in *one* month would be a shaky operation, ut to stretch it for six months is totally unacceptable.

61.85. An applicant for a private pilot's license must have a ninimum of 15 hours solo in an airplane before he can receive is rating. A private pilot is allowed to fly transport aircraft roviding he does not get paid for it. It is perfectly legal for a rivate pilot with 15 hours solo to pilot a high-performance nachine anywhere in the world, just so long as he keeps clear f the clouds.

91.8. No person *may assault, threaten, intimidate or inter-ere with a crew member in the performance of his duties board an aircraft being operated in air commerce.*

121.548. Air carrier inspectors will be *admitted to the ockpit.* A few FAA inspectors are conscientious workers; most f them are disgruntled airline rejects. When they have the op-ortunity to flaunt their power around the cockpit they do so. One such inspector was sitting in the cockpit of a scheduled light and quizzing the captain to such an extent that the captain was visibly disturbed. The trip had just encountered light turbu-ence and the power was naturally being reduced. No sooner were the captain's hands on the throttle when the inspector velled: "For Christ sake take the power off." The captain re-plied, "That's exactly what I am doing."

Cases such as this are many. Instead of sitting quietly, as hey are supposed to do, and observing the flight, all they do is ncrease the potential for disaster by intimidating the crew.

91.11 (b5). All turbojet aircraft [not propeller driven IE prop jets] *will have an altitude warning device installed by February 28, 1972. This device must be capable of giving two signals, oral and visual, excepting when operating below 3,000 feet.* Some sort of warning device had been long overdue. The cockpit of a jetliner can get very busy, and missing an assigned altitude is easy. This regulation has one glaring flaw: if there's a time when a pilot needs help, it is during a low-visibility approach. All possible reminders should be operating when nearing the ground, but that's the time when the FAA sees fit to relax the backup rule and allow for one, rather than two, warnings.

91.70. No aircraft will be operated at a speed in excess of 288 MPH when below an altitude of 10,000 feet.

Many times when approaching a metropolitan airport, ATC may assign an altitude below 10,000 feet, as far as 30 miles from the airport. This puts an economic burden on the airline and delays the passengers. The approximate cost per minute to operate a 747 is $100.

While the airlines must cut their speed below 10,000 feet, fighter pilots are permitted to cruise around at much greater speeds. Since the airways are in a chaotic state, this rule is just a cover-up for years of insufficient planning.

91.25. It is the responsibility of the captain to maintain a continuous listening watch on his ATC radio receiver so that any instructions from the ground controller will be received and acknowledged. It is also mandatory that the individual airlines have a separate communication system of their own, completely divorced from any government radios. This pilot dispatch radio is used continually throughout a flight for such pertinent matters as remaining fuel on board, current-destination weather, and many other important operating aspects.

The more safety-conscious airlines have installed in all their aircraft a radio similar to the telephone in your home. When the company wishes to contact a pilot, the telephone in the cockpit rings. Until that time the crew devote their attention to monitoring the ATC radio receiver only. This phone is not mandatory so only certain airlines are equipped with it; however, a normal radio receiver must be on all the time.

Picture the absolute chaos of a modern jet cockpit: hundreds of dials to watch and continual vigilance outside the cockpit; two radio receivers with both company and FAA blurting their instructions at the same time over the same speakers; add to this the stewardess who has come up front to escape the clutches of a drunk passenger.

And, of course, no man can listen to two radios at the same time and react perfectly to both of them. Yet the FAA says it will be done.

121.105. Each airline must have adequate maintenance and spare parts available at certain points along the routes. This for a change, is an excellent regulation. The only problem is that the FAA does not enforce it. A look at the logbooks of many airlines would show that there are airplanes traveling around with faulty components that haven't been replaced for weeks. The outlying stations claim they don't have the necessary parts in stock, but the discrepancy will be eliminated when the plane goes through the next service check.

Seven out of ten times, when a craft is released for flight after an inspection at the main overhaul shop, the pilot squawk is not repaired.

FAR 91.47 specifies the minimum number of emergency exits for each type of aircraft. The ancient DC-3 (28 passengers) had five emergency exits. The Martin 404 (44 passengers) had seven emergency exits. Along comes the jet age: majestic beauty, tremendous speed, superb cuisine, first-run movies, and less than half the proper emergency exits. A stretched DC-8 has four emergency window exits plus two extra doors, yet this sky queen carries over 250 passengers. One other point: all emergency exits on all commercial airlines should operate alike; once a passenger is briefed on one airplane, it should suffice for any other lines he might fly.

The potential danger of too-few appropriate exits is compounded by the inadequate fire-fighting equipment at many commercial airports. After a pilot calls a tower and declares an emergency he can often crash-land before the city fire engines reach him.

121.443. Scheduled airline captains must make an entry into each airport that they normally fly into at least once a year. Now that's a very fine idea, except for one small detail: it only applies to a scheduled airline captain flying a regular trip. A scheduled airline captain is authorized to fly people to any airport anywhere as long as it is a charter flight. No special airport entry is required. For example, a pilot who for the last ten years has flown in and out of Kennedy Airport as regular as clockwork has his run changed. If he doesn't fly into the JFK area for over a year, he cannot legally land there until he requalifies. He may, however, execute an instrument approach into a mountainous area and land at an airport he has never seen before, just so long as it's not a scheduled flight. It seems that the government is concerned only with *scheduled* safety; charter flights don't count.

121.155. When an airline knows of hazardous runway conditions, it shall suspend operations. A regulation that is seldom complied with. Most airlines are too interested in making money. The reported runway condition should be the prime responsibility of the FAA, not the airlines.

At some airfields the airport operator or the airline's ticket agent reports the state of the runway to the crews. At some fields the airport manager goes out in a car and checks the braking action of the runways. He reports to the tower: "Braking action good." Two airplanes land and both report: "Braking action poor to nil."

Now the economics of the field is threatened; some of the almighty dollars may slip through the hands of the airport operators. When the reported action is near nil, the airport is in danger of being closed. This would mean diverting airplanes to other fields, and the exorbitant landing fees (more than $200 per landing for four-engine jets) would be lost. Sooner than have that happen, the airport man gets in his car again, goes out on the runway, and tries the braking action a second time.

You guessed it: "It's good." Now the jets can land. It may not be safe, but at least it's profitable.

Accurate braking action reports are vitally important to a jet captain. He has to land 200,000 pounds of metal at a speed of over 120 miles per hour. It is imperative that he receive accurate information so that he can plan a safe approach. The U.S. Air Force has long used a small, inexpensive device that is at present the most accurate means for checking braking action. It is called Runway Condition Reading (RCR). The number shown on the pointer on the machine can quickly tell the pilot what the action really is, and not what some unscrupulous brake pusher says it is.

FAR 91.107 states that a pilot in a control zone (usually a five-mile radius from the center of the field) may take off VFR as long as the visibility is one mile and he remains clear of the clouds. A flyer with a student license (perhaps 10 hours' total time in the air) is allowed to take off from a metropolitan airport with a ceiling of 500 and a visibility of one mile. He must report when he leaves the control zone that he is in VFR conditions. Many private pilots (and a few commercial pilots who should know better) use this special VFR to get out of an airport when they lack instrument rating. Special VFR allows a student pilot to land at an airport in VFR conditions?!! While an airline captain approaching the same airport cannot land because the weather is below his limits.

Most commercial airfields have a block of airspace over them called an "air traffic area." This airspace extends from ground level to 3,000 feet and covers a radius of five miles from the center of the airport at which a control tower is operating. Any pilot flying beyond these limits is as free as a bird. The approach zone and climbout paths extend into these supposedly safe areas. Why then is it perfectly legal for an inexperienced pilot to take a joy ride over a busy airport at 3,100 feet and not even tell the tower that he is coming?

91.79. Minimum safe altitude. All pilots will fly high enough so that in case of a power failure they can glide to a safe landing. This regulation is in conflict with the dangerous "special VFR" rules. Should an emergency be encountered, even a light aircraft with a relatively flat glide could probably take off from a metropolitan airport with a ceiling of 500 feet and a visibility of one mile and expect to make a safe emergency landing on the nearest open space—probably Main Street.

At present it is compulsory for commercial turbojets only to fly under positive control at all times (i.e., flights protected by air traffic controllers). All prop and turboprop are excluded from the mandatory regulation. A high-performance military jet is at liberty to fly where he chooses under VFR as long as he remains below 18,000 feet. There are literally thousands of airplanes milling about the sky that are not under any flight plan, and most certainly not under the watchful eye of the controller.

121.163. Aircraft proving tests. Each new type of aircraft purchased by an airline must complete one hundred hours of proving runs over prescribed routes before the passengers can be carried.

When a new model airplane is built it is put through extensive and tortuous tests. In spite of this, the records of approximately 70 percent of all new models are nothing but a series of disasters. Such well-known models as Constellations, DC-6s, Electras, 727s, and Comets, to mention just a few. All of these aircraft had troubles and crashes killing hundreds of people.

Proving flights on these aircraft showed no major design or operating weaknesses. It was not until the models had hundreds of hours in the air that the trouble started. There is no reason why airline passengers should be treated as so many sandbags on a test flight. All major airlines have aircraft in their fleet that carry freight exclusively. Here is an excellent opportunity to test a new model further without jeopardizing the public.

While the criteria used to establish a new minimum figure would be difficult, 100 hours has been dramatically proved to be insufficient.

121.195 (b). When a runway is dry no turbine powered aircraft should be landed unless it can be stopped within 60% of the useable part of the runway.

121.195 (d). When a runway is wet no turbine powered aircraft should be landed unless an additional 15% of the useable part of the runway is available.

121.197 Alternate airport landing requirements are 70% of the useable part of the runway for jet props and 60% for straight jets. Forty-eight hours after this regulation was approved, but not in effect, a Continental Air Lines 707 with sixty-six on board skidded off the end of a rainswept runway at Kansas City and came to rest in a ditch at the airport boundary.

The FAA wet-landing regulation is a step in the right direction, but it is also a cover-up for the lack of adequate runways. According to this regulation an additional 1,050 feet of runway should have been added to the existing 7,000 feet at Kansas City. This would supposedly allow for possible hydroplaning (skidding) on a wet surface. Another alternative to comply with this regulation is to restrict the weight of a plane on takeoff if it is destined for a wet runway on landing.

Oddly enough, the FAA has no 15 percent safety factor for airports listed as alternates. If a runway is wet it is just as dangerous whether it is your first choice or an alternate.

There is no mention made of a restriction on takeoffs. Yet this is one of the most critical phases of flight; it is every bit as important that an airplane be brought to a safe stop if an aborted takeoff is required. On takeoff a plane weighs more than on landing, because it has barely used its fuel. It will therefore require an extra margin of safety to stop—*a margin that the FAA has neglected.*

When is a runway wet? When it isn't dry. Wrong! Only the FAA could come up with a definition of what is wet and what is dry. Advisory circular dated August 17, 1967, states that in order for a runway to be classified as wet certain meteorological conditions must prevail: "Such conditions include showers, heavy drizzle, continuous light rain, moderate or heavy rain, etc." The weather conditions for a runway to be dry are: "Scattered showers in the area, intermittent light rain, light snow, etc."

Next time you are riding with the top down and it starts to drizzle, don't stop to position the roof. You won't get wet, the FAA says so. Any driver knows that it takes only a *damp* highway to cause a skid.

The person most closely associated with the weather at an airport is the tower operator or station weather forecaster. These men should be responsible for the condition of the runways but the majority of them don't even know the federal government's definition of wet and dry. So the airline dispatcher, who may be hundreds of miles from the airport and receives only the reports of a ticket agent, has the all important job of calling the strip wet or dry.

121.471. Flight time limitations. No domestic carrier will allow a pilot to exceed the following maximum flight duty times: 1,000 hours per year, 100 hours per month, one 24-hour period each 7 days free of all duty, a maximum of 30 hours in any consecutive 7 days. A pilot will not be *scheduled* for more than 8 hours flying per day. If, through weather delays, the 8 hours is exceeded, this will be allowed.

Luckily for the traveling public, ALPA realized many years ago that 100 hours per month was too much flying from a safety standpoint so they made the maximum 85 hours per month. This regulation hasn't changed in thirty years, in spite of today's more demanding working conditions.

The FAA regulation requiring one duty-free day in seven is only a token of what it should be. Mechanics, tower operators, dispatchers, all enjoy this luxury. However, there is a difference in the maximum number of hours a dispatcher and tower operator can be on duty as compared to a pilot. The former's working day is over at the end of ten hours; pilots can be on duty indefinitely.

Let's examine this regulation further. A pilot cannot be scheduled for more than eight hours' duty in any twenty-four-hour period. On the surface that may sound like a better deal than the tower man who can get ten hours' duty, but many times during bad weather when the air traffic delays mount up, pilots have been stuck up front for twelve to fourteen hours. The big kicker in this regulation is maximum in *schedule* time and not *actual* time.

Until now we have been discussing actual time in the cockpit. Let's take a look at on duty time. By the very nature of the job, a pilot leads a different life than the nine to fivers. A pilot may be away on a scheduled series for as long as a week. During that time he can fly around the clock every day—as long as he isn't scheduled over eight hours and has eight hours off duty after a scheduled flight series, he's legal.

To drive this point home, take a typical case. Captain Nick Cone has just completed three days of flying which consisted of working during the day and resting in the evening. Now he has two days at home so he decides it's time to start on the recreation room. Nick does not have a regular run. He is a stand-by pilot which means he must be available to fly twenty-four hours a day perhaps six days a week. He is called to fly when the regular pilot is sick or on vacation.

Nick has been busy working on his building project. He has put in an exhausting day and at 11 P.M. he heads for bed. Just as he is starting to doze off the phone rings. It's his airline crew scheduler assigning him a trip that has just been uncovered; the regular captain has called in sick. Nick now has to report to the airport for a 2 A.M. departure. Since it is mandatory that the crew report one hour before flight, Nick will have to get cracking.

Captain Cone's series of flights calls for a round trip between Chicago and New York with four intermediate stops. The weather throughout the entire trip is bad, necessitating ten instrument approaches. Nick's first officer is making his first trip with the airline. Captain Cone will be doing all the flying as any commander will with a brand new co-pilot. The stage is now set.

How would you like to be a passenger on a flight where the flight plan calls for ten instrument approaches, all of them barely within legal limits, with a captain who has not had a wink of sleep in the past twenty-four hours? Not a comforting thought, but it is happening right this minute.

It may sound incredible, but it is true. It is also true that Captain Cone could have refused to take the trip on the grounds that he lacked sufficient sleep, and he would have had a regulation to support him. But the nature of standby flying is such that you know a call may come at any time. Still, it is literally impossible to be completely rested at all times, so you take your chances and so do the passengers.

ALPA is safety-conscious enough to put the maximum on-duty time at fourteen hours. FAA pilots also have a maximum duty time, sixteen hours for a two-pilot crew. Their manual states: "Normally, flight crew members must have 10 hours' rest before the start of any duty period which will require flying." Airline pilots have no such regulation in their FAA manuals. They are allowed a mandatory rest only if they exceed eight hours in one day.

With most people working shorter and shorter hours each day, how is it possible that the FAA puts the simplicity of the DC-3 on a par with the complexities of the 747 and does nothing to change a thirty-year-old regulation?

121.537 Responsibility for operational control of supplemented air carriers (formerly nonskeds) *and commercial operators rests with the captain and the director of operations.*

All scheduled air carriers have dispatchers throughout their systems. A dispatcher must pass a written exam that is very similar to a pilot's. He is licensed by the government and no flight can depart unless the captain and dispatcher agree that it is safe to go. Supplemental air carriers, many of them with equipment as large as 707s and DC-8s, do not have to employ dispatchers. Instead the captain and a nonrated individual called an operations director jointly guide the planning of the flight.

There is no reason why your safety while traveling should be jeopardized, no matter what airline you fly on.

121.575 Alcoholic beverages. No intoxicated passenger will be allowed to board an airliner. Intoxicated passengers will not be served drinks, and no person may drink from his own personal bottle.

This is a good regulation. The enforcement of it by the airlines is something else. More than once I have seen drunks being helped up the aircraft steps by ramp personnel who are only too glad to get rid of the potential troublemakers. This rule should be strictly enforced by the airlines, but to them, every passenger is money, and that's what they are in business for.

The safety of any plane is endangered by drunks. If they aren't chasing the stewardess, they're making general nuisance of themselves to the passengers. Imagine the potential danger a drunk passenger can cause during an emergency landing. He won't have the foggiest idea of where the emergency exits are located, and may hinder other passengers who are trying to evacuate a burning plane.

121.687 (b). Dispatch release. The release from a dispatcher for every flight must contain a copy of the weather report for all stops anticipated by the carrier. This report must be the latest available. If the teletype weather machine at an airport becomes inoperative and no new weather sequences have been transmitted for some time, it is perfectly legal to use a weather report from last week, as long as it is "latest available."

Sports parachute jumping: the FAA authorizes sky diving on and off the airways. There are numerous jump centers throughout the United States that have their main base of operations directly under busy jet airways. The jump ship must be equipped with a two-way radio and the pilot must inform the nearest FAA facility five minutes before he drops the jumpers. Many delayed jumps are started above 12,000 feet. At 10,000 feet and above, aircraft can fly as fast as they choose. Imagine a jet traveling at close to the speed of sound as it plows into a sky diver.

Each scheduled carrier has a separate section in its manual of compulsory federal regulations called Operations Specification. These regulations may or may not apply to other operators, but they are mandatory for the carrier to which they are directed.

12. Special provisions turbojet aircraft. Turbojet aircraft will be flown under IFR at all times.

There is no reason why the captain of a turbojet is required to fly under the watchful eyes of the radar controller while prop pilots fly helter-skelter under no one's positive control. Prop passengers are entitled to controlled safety as much as anyone. Even though a pilot can file an IFR flight plan at any time, there is no regulation making positive control mandatory for all air carriers. And there are still a lot of propeller-driven airliners around. In 1968 only one-half of all the airlines' fleets were turbojets.

FAA *Air Carrier Operations Handbook* has an amazing statement: no four-engine turbojet will operate into any U.S. airport that lacks a control tower. Operations outside the United States can be approved.

Two- and three-engine turbojets do not need a control tower to operate. I see no reason why the passengers on a small jet are not afforded the same amount of safety as those on a large one. What has the size and same relative speed got to do with whether or not an airport has a tower?

The Federal handbook entitled *Operation of FAA Aircraft* has some interesting statements:

Chapter 2, paragraph 11D, requires FAA pilots to have a minimum of 48 hours of pilot time during the previous 12 months. There is no such rule for civilian flyers who are legal with three landings in 90 days.

Chapter 2, paragraph 20, relates that all FAA pilots who fly turbojets above 25,000 feet are required to obtain physiological training. This training consists of study, lectures, and a ride in the pressure chamber (a pressurized tank similar to an aircraft fuselage). Here various emergencies are simulated, e.g., explosive decompression similar to a bomb explosion on board, hypoxic symptoms (oxygen starvations), etc. No airline crews are given this valuable training.

Chapter 4, paragraphs 34–42B. Fueling operations: Be sure engines are not running and that no electrical switches are activated. This is a good idea; too bad there is no similar regulation for airlines. Often jets with faulty engine starters are kept running all day even while fueling is in progress. It would seem from the above that the FAA takes better care of its own than it does of the traveling public.

According to FAA statistics the number of passengers carried in 1969 was over 120 million. This figure is forecast to reach 379 million by 1975.

Before the FAA changes a regulation they issue a docket to all interested parties stating precisely why they want the particular change made. There are usually a few months allotted for the various pros and cons to be submitted. If after a meeting of the parties involved a change is decided upon, it may not be implemented for months—in some cases, years.

The ATA should not be allowed to wield so much power. If a proposed rule will cost the airline some additional money they scream—until the FAA cries uncle and backs off.

If only the Federal Aviation Administration would live up to its latest advertisement: *Safety in the air is the primary mission of the FAA.*

Chapter Seventeen

JOHN'S FLYING CLUB

MANY GOVERNING bodies in Washington know their limitations—the Federal Aviation Administration is not one of them. FAA Administrator John H. Shaffer, who rules a vast empire of 48,000 supposedly safety-conscientious employees, is one of the worst offenders.

Since Shaffer came to power in 1969 he has done little other than talk. If it were not for the PATCO "sick-out" in March 1970 he would probably have served his term of office and faded quietly away as one more do-nothing administrator. During the dispute between the controllers and the federal government John Shaffer used every rule in the book to put the controllers in a bad light. His subordinates lied about the number of controllers on duty, telling the public there was no cause for alarm, the airways were as safe as ever. His daily controller count included all employees of the buildings: mechanics, janitors, painters, and drivers.

Mr. Shaffer has taken over an agency created by Elwood R.

Quesada when he was elected chief of the newly formed FAA in 1958. Quesada supervised a vast fleet of aircraft and the crews to man them. He spent millions of dollars to train pilots who would ride herd on the airlines. The air carrier inspectors are still with us and their numbers are increasing.

The FAA has cried poverty for years. Never enough money for proper airports or landing aids, but always sufficient funds to support an FAA fleet of over a hundred aircraft, larger than many airlines.

Let's examine the job requirements for the men who can make or break an airline captain. An air carrier inspector must be at least twenty-three years old and possess an air transport rating which requires a prerequisite of 1,500 hours' command time. (There are very few airline captains who don't have many thousands of hours, and not in light airplanes.) Perhaps a handful of men actually wanted to be air carrier inspectors. The majority of them couldn't get an airline job or failed to meet the airlines' high standards and were fired. It is a known fact that an airline pilot's job is the goal of all who fly for a living.

Airline captains pass two full proficiency checks a year as well as recurrent ground school tests and a route check. An FAA air carrier inspector functions in his position with less testing than the man he is checking. Inspectors are required to have two proficiency checks a year, but under certain conditions they may waive one of them. The air carrier inspectors' handbook has this to say concerning flight checks:

> Waiver authority. The region may authorize inspectors to conduct airman certificate work when inspectors have not had recurrent training: provided training has been properly requested and courses are available, or when an inspector is unable to attend the assigned course due to sickness.

In accordance with their manual, inspectors must receive flight training once a year at their main flight operations center in Oklahoma City. No two airplanes fly or look exactly alike. Each model has different instrumentation, emergency procedures, capacities, weight and flight characteristics. An airline pilot is required to go to extensive ground school and flight training for each new piece of equipment he flies.

The FAA does have an extensive fleet of aircraft, but they do not have every type of plane in airline service. What happens when an inspector is due for his annual proficiency check and he is assigned to an airline that flies DC-8s? He takes his check ride on a DC-9. The only similarity between these planes is the fact that they are built by Douglas. The DC-8 is a large four-engine transport weighing 350,000 lbs. The DC-9 has two rear-mounted engines and is considerably smaller, weighing only 90,000 pounds. Since usually the only flying an FAA inspector does is when he gets his proficiency check once a year, it behooves him to get it in the right piece of equipment.

It is conceivable for an inspector to get his initial checkout in an airplane that he may never again fly. Yet this man has the authority to ground a pilot who has been passing proficiency checks for years, and who may have 15,000 hours in an aircraft that the inspector has been in for as little as 15 hours. Even a highly skilled airline pilot would not attempt to question the operation of another pilot on a plane in which he was not experienced. But an FAA inspector is granted this power by government decree.

Here are a few of the aircraft that the FAA says are similar enough for an inspector to train on: BAC-111 and Caravelle are the equivalent of a DC-9, the Constellation is similar to a DC-6, a Viscount and an Electra are one and the same. Well, at least the last two are turboprops, but that's where the similarity ends.

Many inspectors do their job in a professional manner. A few of them are so frustrated because they could not make the grade with an airline that they have nothing but envy for those they are testing.

There isn't a pilot in the sky today who doesn't break at least one regulation every time he flies. It is practically impossible to comply with one regulation without infringing on another. Any inspector who is out to get you most certainly will.

And he is the law; there is nothing to do but wait until your case is heard.

Inspectors have caused undue hardship for many pilots. In numerous cases their very presence in the cockpit has nearly caused disaster. One inspector was riding the jump seat and was quizzing the crew so diligently that they nearly had a mid-air. Another captain was so distracted by the questions put to him that he tuned the wrong ILS in and questioned the tower as to its operation. The following is a classic example of how one of the government's representatives conducted himself on a recent flight.

Before the trip had started, while the crew was making the flight plan in Operations, this inspector let it be known that he was in a foul mood. He made statements to the effect that he was going to bust the next captain he rode with—and that was no idle promise.

During the flight he conducted a quiz composed mainly of "nitpicking" questions. Inspectors always have an unending flow of knowledge about insignificant tripe. They diligently study a manual concerning a minor component in the aircraft and then proceed to flaunt their knowledge about it just to show the captain who's boss.

All jets have a speed-warning bell attached to the airspeed indicator. Whenever either of the two needles exceeds a given speed, the warning bell sounds. If the two airspeed indicators are not even, the one that reaches the limit first rings the bell. In the case we are discussing the co-pilot's indicator was a few knots faster than the captain's and it triggered the warning.

This was the cue for a flow of foul language from the inspector.

The captain was already in the process of reducing power. There was absolutely no danger in the fact that this bell was ringing just so long as it was not ringing for an extended period.

Here you have a picture of how an inspector can make the cockpit a perilous place. With a three-man crew, at least there is a third person to check the instruments. But with only a captain and co-pilot the stage is set for disaster. By now the crew were so on edge they began making foolish mistakes. When the flight landed at Fort Lauderdale the inspector mumbled something about calling the dispatcher and stopping the flight. When questioned by the captain, the inspector would only say, "You rang the bell. That's a violation."

The examiner would not commit himself as to the disposition of the flight. When approached by the co-pilot he snarled, "You keep out of this or I'll violate you too."

The FAA inspector finally decided he would take the captain off the flight and ground him for a two-week period and demand that he be given additional training. The anxiety the captain was subjected to was uncalled for, not to mention a $1,500 pay loss.

Since 1963 the average number of violations against pilots has remained constant, about 250 a year. Most violations carried a penalty of three to four weeks' pay which is in the thousands of dollars. Here are a few of the violations and fines incurred.

A captain allowed a passenger in the cabin to occupy the stewardess's jump seat: 60 days' suspension. Another pilot took off with the gear pins installed (a pin that keeps the wheels from collapsing on the ground): he got two weeks. A Braniff pilot flew his approach to JFK lower than prescribed: he was demoted to co-pilot for six months which cost him $11,000.

Many pilots are coerced into accepting penalties they are not guilty of. The chain of events concerning a violation follows:

1. A telephone or personal interview is conducted by an investigator.
2. A certified letter relates the facts concerning the violation and a statement is made to the effect that the pilot is subject to a $1,000 fine for each violation. A compromise settlement is also usually mentioned.
3. At this point the violator can answer by mail, stating if he is agreeable to the charges and the fine. If he accepts the

case is closed. However, the airman's files will have a permanent copy of the charges.

4. If the pilot pleads not guilty, the case is turned over to a U.S. attorney and a federal court takes over. There have been only fourteen civil court cases settled since 1926.

The threat of a $1,000 fine as compared to a $300 compromise is a powerful force. And when a pilot has to pay at least a $500 attorney's fee for federal court plus numerous other expenses, he really has little choice. It is costly to buck the government; that is why the FAA gets so many flyers to agree to the false charges against them. The system that gives a man "sentence first, verdict afterward" is contrary to our fundamental principles.

On September 15, 1966, all turbine-powered aircraft had to be equipped with a cockpit voice recorder. This was to be used by the investigating team to aid in solving the cause of an accident. It would, however, not be used to violate a pilot. To date at least one airline has admitted it terminated a pilot solely on what was found on his tape. As this action received widespread coverage among flight crews, some pilots now turn the recorder off, thereby eliminating a contrivance that might lead to dismissal. This defeats the stated purpose of the device, a valuable aid to accident investigation. But it does take two to play the game.

Section 314(a) of the 1958 Federal Aviation Act "permits the Administrator to designate private persons to act for him in various capacities, among them the testing of applicants for airman certificates." This allows fully qualified airline pilots with years of experience to do the government's work for them and at no cost to the taxpayer. In order for a designee to become qualified he must conduct a complete check ride while under the watchful eye of the FAA agent. As soon as he receives his rating he immediately saves the government a minimum of $12,000 per year, the lowest salary level for FAA inspectors.

When an airline pilot receives his final check ride there are usually two other men in the cockpit with him: a company-designated check pilot and an FAA air carrier inspector. The airline pilots occupy the left and right seats, while the inspector rides in the jump seat. Every pilot who upgrades must satisfy both the company check pilot and the FAA. There is no conflict-of-interest situation here, because no *line* check pilot will ever pass a man unless he cuts the mustard. Having the relatively inexperienced FAA pilot on board only adds to the already high level of pressure in the cockpit.

How can a man who has never been an airline pilot, who flies only 20 hours a year, and who has one-eighth the experience of his students ever hope to equal the competence of an airline pilot?

There is an FAA stipulation about airline instructors. They cannot examine the pilots they have trained. The reverse is true concerning the government inspectors: the FAA rules state that a federal inspector may only be tested by another FAA examiner. John Shaffer doesn't want the flying club section of his empire disturbed, so instead of increasing the ranks of the airline pilot examiners, he is seeking to enlarge the FAA pilot group.

On April 7, 1970, Representative Edward P. Boland of Massachusetts, chairman of the Subcommittee on Appropriations, questioned Mr. George Moore, associate administrator of Operations for the FAA.

"Mr. Boland, how many airplane pilots do you have to operate your aircraft?"

"Five hundred twenty-three."

"Mr. Boland, would you provide for the record the estimated total salary cost for these pilots in fiscal years 1970 and 1971."

"The estimated total salary cost for 523 agency pilots is $8,348,000 in fiscal year 1970 and $8,486,000 in fiscal year 1971."

By 1981 the FAA plans to have spent in the neighborhood of $123 million for aircraft and flight training.

On January 6, 1971, a Federal Aviation Administration Douglas DC-3 crashed while executing an instrument approach at LaGuardia Airport, New York. At Prospect intersection, LaGuardia approach control requested the pilot to change over to tower frequency. This was not done for over four minutes; the co-pilot was having difficulty finding the proper frequency. Twelve seconds before the crash the co-pilot looked out to see the approach and runway lights ahead. He thought they were too low but "if they didn't encounter any obstacles" they might make it.

They did encounter a few obstacles. They flipped over on their back in a residential area 2,000 feet short of the runway. The LaGuardia weather was bad but well within limits for an approach. In fact, two minutes prior to the crash, an air carrier Boeing 727 made an uneventful approach and landing. Two minutes after the accident another Boeing 727 airliner shot an approach with no problems whatever. The runway in question has a full ILS system with glide-slope approach lights, high-intensity runway lights, sequence flashers, and an operating control tower. The entire approach was poor and the copilot's own statement proves this. He said that when they broke out of the clouds their ILS localizer was showing them off the centerline by two-thirds deflection. They were also high on the glide slope. Any pilot who is that far off that close to the runway would have been wiser to make a missed approach. Both pilots had physical waivers. They were required to wear glasses and the captain had a hearing problem. In spite of the vision difficulties and the rain, the windshield wipers were never turned on.

The NTSB blamed the crash on poor cockpit coordination and bad procedures. These conditions resulted in a mismanaged ILS and landing approach and the continuation of the descent into ground obstructions.

It is interesting to note the different slant an NTSB investigator has when working on an FAA crash vs. an air carrier. Long before the final decision was drafted, the head investigator was quoted in the press as saying that there was a weather phenomenon on the night of the crash which was hard to detect. He said that as the DC-3 was about to land the wind shifted 180 degrees, creating a "hazardous" condition for landing. Yet airliners landed before and after the FAA crash with no problems.

The FAA is asking for an additional $549,000 in order to pay for out-of-agency pilot training. Seventeen specialists will be trained in the 747 and two in the DC-10.

At Oklahoma City the ramp is chock-full of expensive aircraft. Ninety-one are owned outright and 11 are leased, a total of 102 planes that cost in the neighborhood of $50 million.

The agency does need aircraft for research and development, and for flight inspection, but when it comes to training air carrier inspectors, that's another story. Aircraft such as the Boeing 707, Douglas DC-9, Lockheed Electra, Douglas DC-6, Boeing 727, Convair 340, and Douglas DC-3 are all there waiting to burn up needless fuel to train unnecessary pilots. In fiscal year 1971 the total cost to the taxpayer for the flight crews, mechanics, and program director was $42,703,000. There are numerous FAA business jets that are plushed up for executive use. It is safer and cheaper to ride on a commercial airline.

Senator Vance Hartke of Indiana has called for a full-scale investigation of the FAA. He charged that the agency could prevent most of the accidents that have claimed so many lives. He said that the FAA has a "*pitiful lack of concern for Aviation Safety.*"

The flying business should be left to professional full-time experienced pilots. They should be the ones designated by the FAA to give the checks. All of the general aviation instructors are civilians and are not FAA employees. Airline pilots should also be afforded an honest check by a man who knows what he is doing.

Money can save lives. The sinful waste of millions of dollars by the FAA on aircraft and pilots would buy the necessary landing aids to prevent the next disaster.

In 1974 Congress allowed the FAA to purchase seventeen new aircraft at a cost of many millions of dollars. Now the Feds are asking for $3,100,000 to purchase a DC-10 simulator. This is to train their pilots to check out and oversee future DC-10 airline pilots. All this is entirely unnecessary because the airlines still do all their own pilot training and checking. The FAA only issues jet ratings after the applicant is fully trained by his company.

In the days of propeller airplanes, the company check pilot was allowed to train and issue an FAA license without an FAA inspector ever taking part in any of the training. Since the introduction of the jet, however, the airline check pilot is allowed to fully train a pilot and when the pilot is ready for his rating ride, the FAA takes over and issues the license. I believe that is called job security.

I have no objection to the FAA riding with me; my complaint is the tremendous amount of money it takes to train such personnel. In 1974 the FAA pilots flew 11,484 hours at a cost to the taxpayer of $3,500,000. The total fleet consists of ninety-two aircraft, and salaries of the pilots who flew them in 1974 amounted to $5,532,200. The cost of fuel has increased 59 percent since 1973, so the FAA fuel bill for 1974 showed an increase of $1,487,000 for a total of $5,629,357.

Most of the millions of dollars spent to train the 400 FAA inspectors is wasted money that should be spent on installing ILS. The FAA always cries poverty, but the money saved by cutting back their flight training would pay for many ILS, the approach aid that is usually installed after the crash.

On March 25, 1975, Alexander P. Butterfield resigned as FAA Administrator. Of all the past administrators Butterfield was the best. Although he had many bureaucrats to contend with, I feel sure he did the best he was allowed to do. His aviation career was outstanding and they will search far and wide before they find a more competent man for the job.

If the FAA is ever going to fulfill its assigned task, it should be divorced from the Department of Transportation, and the game of politics should be left to the politicians.

Chapter Eighteen

THAT OTHERS MAY LIVE

EVERY PILOT has a burning desire to upgrade into bigger, faster, and more productive equipment. Many wait years for their chance because of the democratic seniority system.

Captain James W. Morton, forty-eight, was originally hired by Chicago and Southern Airlines (C&S merged with Delta in March 1959). In July 1959, after eight years as a co-pilot, he finally got his chance to upgrade on a four-engine jet. Jim was a highly skilled pilot with over 16,000 hours in the air. His license proudly displayed the fact that he was captain-qualified in the DC-3, DC-6, DC-7, CV-240/340/440, and the DC-9. His latest proficiency check was conducted in the DC-9 in September 1966.

The DC-8 was the next plane that he had to master. Jim had already successfully completed all phases of ground school and simulator training and was presently undergoing his final grooming at New Orleans, Louisiana before he would take his FAA rating ride. Captain Morton was an exceptionally apt student but because of the voluminous amount of knowledge that must be absorbed during flight and briefing periods he decided to take a room at a motel near the airport where he could devote all of his time to this latest challenge.

With fourteen hours of intensive flight training under his belt, he felt he was ready to tackle the FAA oral exam. On March 28, after attending eight hours of final ground school briefing, he took the equipment exam. He was asked a multitude of questions, most of which had little bearing on actual day-to-day operation of the DC-8. He had memorized what he thought would be on the quiz because to understand the intracacies of complicated circuitous diagrams completely would be possible only for a skilled aeronautical engineer.

Jim was asked: What is the total weight of the aircraft? How high is the tail? How long are the wings? What is the tire pressure? (Mechanics take care of this item.) How many vortex generators on each wing? (This is of little concern to a pilot since the only check he has to make is to be sure that none are broken.) What is the RPM of the APU? (There is no RPM indicator in the cockpit.) At what temperature do the various overtemperature lights come on? (The fact that they light up is sufficient. It tells you that it is time to do something.) On and on the irrelevant questions continued. But Jim had all the answers. To fail this exam would mean a trip back to ground school and the possible termination of flight training until a later date.

After eight hours of ground school, and a few more absorbed in FAA trivia, Jim was ready for a good night's sleep. He was not to get it. After going to his room and studying until about 9 P.M. he turned in, his brain a mass of temperatures, pressures, and procedures. After a two-hour nap the alarm rang; it was time for flight training which would begin at 11 P.M. Practically all of Captain Morton's flight training was conducted at night because that was when a spare aircraft was available.

After a two-hour flight he returned to his motel—dog tired

—and slept for five hours. He attended another full day of ground school and at 6 P.M. went to his room for a rest. He was beginning to wonder if he would ever get a full night's sleep. At 11 P.M. he was once again preparing to strap the DC-8 to his backside and continue flight training.

March 29. Jim reported to the field for his fifth training hop with Captain Maurice Watson the company check pilot.

"Hi, Maurie," Jim said. "How's everything going?"

"Well, I had a pretty good day today. Played golf most of the morning. Shot an eighty-five."

"Say, you're really getting down there. Well, my day consisted of another filled with mock-up, pressures, and temperatures."

"You're a glutton for punishment, Jim. We'll take it easy tonight. Just a few emergency procedures and then we'll spend the rest of the training period shooting ILSs. The FAA will be with us tonight."

The preflight check was completed and a clearance to runway 1 was received and acknowledged. This runway has no ILS for the direction in use, and is marginal in length, only 7,000 feet.

The first simulated engine failure occured at V1. As the plane was now committed for takeoff, Jim kept his heading down the runway and waited for the next call from the check pilot. "Rotate," said the instructor. Captain Morton skillfully applied back pressure to the stick and 180,000 pounds of dead weight reached for the evening sky.

One minute later check pilot Watson chopped a second engine. The craft had now lost the nos. 1 and 2 engines which meant it would have to be nursed along with only 50 percent of its total power in use. The altitude was 1,200 feet and the airspeed 200 knots. At a critical time such as this, any airspeed lost can be recovered only by pushing the nose down. This would sacrifice precious altitude that could not be regained.

Jim called for the appropriate emergency check list and the engineer started to clean up the two dead engines (i.e., perform the necessary check items). The check pilot signified to Jim that he had just lost his rudder boost. (This was simulated by pressing the warning light.) One more emergency to contend with. As the flight turned left the altitude was down to 900 feet.

Jim called "Flaps 25." As the flaps were sliding down their tracks, a feeling of being thrust forward in the seat signified that they were indeed doing their job.

The check pilot began to prompt his student: "Don't get below 160. . . . Ball in the middle whatever it takes to put her in there now. . . ." (On most airplanes the needle and ball instrument is nearly hidden behind the stick.) So far everything was going as advertised. The landing check list was called for and the check pilot elected to dump fuel flaps. Shortly thereafter, the airspeed bled off to 165 knots. They were now 2.5 miles away from the runway and without the benefit of any glide slope, either visual or electronic, it would be a game of chance to judge the proper approach profile.

Check pilot Watson finished the last item on the check list and stated, "Before landing complete, 129 knots is approach, 24 is threshold. . . . Okay, looks good. How 'bout that now, we're straightened out."

The lights of the runway were now only a mile away and closing rapidly.

"Call my airspeed for me," Jim said with authority.

"140," the instructor replied.

From this point the need for immediate corrective action became apparent. The airspeed was dwindling to an unsafe condition. At 136 knots and half a mile from the strip, the need for more thrust was readily recognized. Power was applied to Nos. 3 and 4 and the plane made an immediate turn to the left coincident with a drop in airspeed and altitude.

"135 is your airspeed," shouted the check pilot. "See you're letting her . . . put the rudder in there . . . you're getting your speed down now. . . . You're not going to be able to get it."

The ground was now only seconds away.

"I can't hold it, Bud," exclaimed Jim.

Captain Watson shouted, "NAW, DON'T. LET IT UP . . . LET IT UP . . . LET IT UP!"

The cockpit voice recorder ended at 12:50 P.M. And one of man's most magnificent achievements fell to earth instantly, killing the five men in the cockpit. As it knifed through two forty-foot trees it continued its destroying journey, clipping a corner off a house. It lost some of its deadly momentum as it plowed through another house. A third dwelling was demolished, and as the fiery missile's speed diminished it skidded over a railroad embankment and came to rest against the Hilton Inn Motel. Here the remainder of the hulk was demolished by explosion. In its 700-foot wake thirteen more people lay dead.

This training catastrophe is only one of a long line of needless deaths. The following list compiled by the Flight Safety Foundation deals exclusively with engine-out maneuvers. If a worldwide list were included it would fill many more pages.

2/25/59	B-707	Pilot failed to maintain flying speed during two engine minimum-control speed.
1/26/61	B-707	Due to loss of control A/C dove into the ocean.
9/13/65	CV-880	Loss of control during simulated engine-out takeoff.
9/29/66	B-720	During a simulated engine flameout student misjudged distance and undershot runway.
2/20/67	DC-8	Aircraft crashed during simulated engine-out.
3/30/67	DC-8	During two-engine-out landing aircraft crashed.
12/30/68	DC-8	Loss of control during simulated engine-out.
7/26/69	B-707	Aircraft went below minimum con-

		trol speed during a two-engine-out maneuver.
6/24/69	CV-880	During critical engine chop on takeoff aircraft was destroyed.
3/31/71	B-707-720	Aircraft crashed during three-engine abort.

Statistics prove conclusively that of all the fatal training accidents occurring in this country, over 70 percent are caused by engine-out maneuvers, 17 percent in simulated hydraulic emergencies, and the remaining 13 percent under flight control conditions.

All pilots have to go through the dangerous engine-out maneuvers. The main difference between training on a four-engine vs. a twin-engine jet is the fact that only on initial upgrading is a four-engine pilot required to demonstrate his ability to fly his machine with half the power gone. A pilot whose plane only has two engines is required to fly with half the power gone on every check ride.

On February 25, 1959, a Pan American 707 check pilot was demonstrating controllability at minimum speed. The airplane went through two complete turns of a spin the force of which detached the No. 4 engine from the airframe.

The Air Line Pilots Association has for years requested that dangerous maneuvers be performed in flight simulators while safely on the ground. Today's simulators are so amazingly realistic that at times you have to remember that you are not in the air. A two-engine-out landing is an emergency situation; the FAA should not treat it in the same manner as an ILS or a normal landing. As there has never been an accident caused by 50 percent power loss in airline operation, it is a rather useless requirement anyway.

On a recent training flight the flight crew was practicing an engine-out missed approach when they were confronted with a real hydraulic emergency. The company check list was executed to the letter but the plane crashed and killed the five crew members. After the accident the manual was changed. Flying near the ground with engines cut is extremely hazardous in itself. Add to this a real emergency and there may not be time to cope with it.

On March 31, 1971, Western Air Lines Flight 366, a routine check flight, departed Los Angeles Airport for a trip to Ontario, California, where some practice approaches would be performed. The only occupants of the four-engine Boeing were the crew plus two jump-seat pilots. The flight was routine until approaching Ontario where approach control was contacted. A clearance to shoot an ILS was received and No. 4 engine thrust lever was reduced to simulate the loss of an engine. The approach was normal, and as the plane passed the middle marker (approximately one-half mile from the runway) the check pilot instructed the pilot to execute a missed approach.

Power was applied to the three good engines and a pull-up was started. The landing gear was retracted, and as the turbines screamed to full life again the popping sounds of compression stalls were heard. The craft pitched up at a steep angle and rose to an altitude of 500 feet. Here it rolled to the right and nosed into the ground in a vertical position. Five more pilots sacrificed on the training altar.

The official cause of the accident has not yet been found, but the NTSB did find a glaring discrepancy in the aircraft. The rudder-actuator support fitting was in failed condition. This resulted in a complete loss of rudder control (the most important control to maintain direction). The rudder support fitting was taken to the Boeing plant in Washington. Numerous stress corrosion cracks were discovered to have existed prior to its failure. The airline had properly inspected the lugs within the time allotted by the FAA. Obviously the time between inspections was too long.

The NTSB stated that it had been advised of similar cases in which both upper and lower rudder-actuated support fittings had failed. In each case the failures occurred during the high stress placed on them during engine-out pilot training maneuvers. These aircraft are the same ones used in regular passenger service. Airlines seldom keep special planes aside for training only.

In recent years at the persistence of ALPA, the FAA has altered its engine-out training requirements. They have waived the engine-out portion for subsequent check rides. But it still must be performed for initial upgrading. One step forward, but not the whole answer. This step simply reduces the frequency of crashes; it will not eliminate them.

The U.S. Air Force forbids its training pilots to cut more than one engine below 10,000 feet. One throttle may be retarded but not lower than 500 feet above the terrain. "Two-engine-out landings will not be practiced," says the military manual.

Clay Vagneur, United Air Lines check captain and ALPA training chairman, had this to say at a recent Air Safety Forum: "Mandatory or planned frequent exposures to simulated engine or engines inoperative maneuvers under all kinds of conditions will eventually result in accidents."

The whole concept of check rides today is wrong. The tense "You'd better do it right or you've had it" attitude of the FAA should not be permitted. Checks should be instructional rides, with training the dominating factor.

Even today the check ride is not as it should be. It still is not a training session. Granted, if you make a mistake, the check pilot will probably give you another shot at it, but it had better be right this time.

In 1973 the FAA changed many of its training regulations. It is now possible to complete nearly all the prescribed checks in the simulator. Many of these simulators are so realistic that you feel you are really airborne. The cockpit is an exact replica of the plane it represents. The check pilot sits behind the crew where he can perform many tricks. He can simulate rough air, cross winds, fog, failed engines, faulty hydraulics and pneumatics, overspeed turbines, fuel leaks, control problems, and many more.

There is an unwritten law that is enforced by many of the better airlines regarding system failures. Up until a few years ago an FAA inspector could pile on the malfunctions one after the other. Now you get one at a time.

Even though the simulator looks and sounds like a real plane, it does not fly like one. The control pressures are different and on takeoff roll some of them resemble an accident waiting to happen. These simulators should be used to check pilots on procedures and emergencies. Until they are refined, the actual flying part is far simpler in a real plane.

When a pilot flying the "box" (as it is affectionately called) crashes, he will only have hurt his pride, not a $25-million airplane.

Because of the energy crunch, simulators are used almost exclusively. One airline estimates that 25,561,115 gallons of fuel were saved in 1973. Operating the boxes costs about one-tenth what it costs to keep a plane in the sky.

The simulator is a far more realistic piece of equipment than the plane when it comes to emergencies. For example, in a real aircraft you can't simulate low fuel contents. There is also no way you can produce icing conditions at the flick of a switch. When two engines are cut during flight on the DC-10 or L-1011, power is also reduced on the third engine to counteract the thrust being generated by the other two. The result is not nearly as realistic or as safe as the box. Here the engines are completely shut down.

The only time a real aircraft must be used for training is when a pilot is up-grading. Even then over 90 percent of the check is performed in the simulator. Today, the pilot who has to take two checks a year can take them both in the safety of the box. He will be required to perform two normal landings in the plane, but this can be accomplished on a regular flight under the watchful eyes of the check pilot.

Obviously, we are moving forward in the area of training. If only the concept of "do it right or else" would change to "keep doing it until you get it right." Only then will the stigma of busting a check ride be gone forever.

Chapter Nineteen

DON'T ALWAYS BLAME THE PILOT

WE IN AMERICA are sometimes overconfident about the planes certified in this country. United States airplanes are usually better constructed, but the criteria for certification leave much to be desired. Many times, seemingly insignificant items go undetected until a tragic crash thrusts them into prominence.

The FAA manual of certification of new aircraft is sadly lacking in up-to-date methods. When the French and the British require a specified test it is spelled out in precise steps. The FAA merely states what is expected on a given test, and the methods leave the manufacturer and their test pilots almost totally in command. Many aircraft companies have their own test pilots who are FAA designees. This means that a pilot in the employ of an aircraft manufacturer is at liberty to "change hats" and pass final judgment on the craft he has been testing for months. Granted he knows the plane better than a full-time FAA inspector, but there most certainly is a conflict of interest involved here. Few men will reject a plane built by their employer.

What follows is some dialogue taken from the Congressional Aviation Subcommittee hearing for 1971. Mr. Boland is questioning Mr. Moore of the FAA.

BOLAND: What are FAA procedures with respect to certification of aircraft?

MOORE: FAA has resident engineers based at the domicile of the manufacturers. The FAA engineers take part in the production and type certification of the aircraft.

BOLAND: To what extent do you delegate to the manufacturers the authority to approve the safety of an aircraft?

MOORE: On large aircraft we delegate about 70 percent of the work in the type certification, and in small aircraft we do a lot more than that. Actually it goes up as high as 90 percent because of the magnitude of the activity and the limited amount of people that we can provide.

Let's take a look at just one of the numerous tests a new plane has to pass. Practically all tests performed before a plane gets certified were conceived long before the jets arrived on the scene. Not only are the day-to-day flight regulations archaic, but the methods of certification need immediate modernization.

Once a pilot has started his takeoff roll and an emergency arises requiring an immediate deceleration, will there be sufficient runway left for a safe stop? This measurement is known as the accelerated stop distance. During takeoff there is a definite point on the runway beyond which the flight cannot be stopped. Prior to reaching this point, the plane can be safely stopped, but once past it, the aircraft is committed to flight. How does a pilot know when he passes this point? His determination depends on the imperfect parameter speed, rather than distance. This highly crucial commit point is called V1 (velocity 1).

The V1 concept presupposes the sudden failure of an engine, as this was believed to be the most critical malfunction at the time the regulation was written. Under the present performance concept the speed V1 will not be reached until the craft has already passed the point where insufficient distance remains for the average airline pilot to stop his craft safely.

Service history indicates that the assumptions derived for certification have not been realistic. We continue to have catastrophic overrun crashes, even when emergencies occur at substantially lower speeds than V1.

Let's look now at some of the factors that render key certification assumptions invalid.

Runway positioning. In current airline dispatch procedures the full length of the runway is always figured. In reality this length may be reduced by as much as several hundred feet. It is virtually impossible to taxi onto a runway and have the main wheels positioned at the very end. Both the British and Australian governments take this positioning factor into consideration. The U.S. Air Force eliminates 200 feet from all runway lengths for four-engine jets.

Factors tending to degrade acceleration. Weights and balance are many times in error. Some planes are overloaded by as much as 10,000 pounds.

Runway surface. No tests have been conducted to determine the detrimental effect of rough runways. All criteria are based on a clean dry runway. In reality this seldom exists. Runway gradient figures are almost impossible to come by, and the temperature and wind reading taken at most airports is at times more than a mile from the actual runway.

Transport engines get a considerable amount of use and are not as efficient as the new ones used during the tests, a far cry from those used daily on airliners. When the captain decides to abort his takeoff, how long should the whole process take? Less than 3.5 seconds, says the FAA.

First the pilot must recognize that a serious failure has occurred. If he is approaching V1 speed he may have only a very few seconds to decide to continue his roll or to stop. If he decides to stop he must apply his foot brakes, retard the throttles, and raise the spoilers. He must continue to keep the craft on the runway by using the nose steering. All these actions are necessary to stop the continued acceleration and the wasting of precious runway—and all must be done in under four seconds. Human time lag tends to be reduced with practice. It is not a true test to expect an airline pilot, the man for whom the craft is being tested, to react as rapidly as a well-rehearsed test pilot. An airline pilot may never be required to stop his aircraft at V1 until a real emergency arises. The certifying test pilots are well schooled in the stop tests, and can perform them flawlessly.

The FAA has thrown another hooker into these tests. The sequence that must be utilized to stop the craft is entirely foreign to the method used after each landing. A normal landing is terminated by closing the throttles first, rather than applying brakes first, as was required for the tests. Certification should be made more realistic and should consist of the usual method of deceleration that a pilot performs after every landing. Because, crisis or no crisis, that is the way he will stop his craft.

The French allow two seconds between each control movement. The U.S. Air Force KC-135, the military counterpart to the Boeing 707, allows a six-second transition from the point of failure to application of brakes. This gives the pilot time to recognize the situation and make a correct decision.

The manufacturer has proved that his test pilots initiated abortive action immediately at critical engine speed, and the aircraft forward motion only increased a single knot. A controlled test run in an appropriate aircraft simulator proved conclusively that the certification tests were not valid. Forty-two experienced airline pilots evaluated the accelerated stop tests. For all runs, aircraft velocity increased from emergency to peak speed from 4.9 to 15.2 knots with an average of 9.9. These values disprove one manufacturer's assumption that only one knot of overspeed was required for an abort commencing at V1. Total time consumed to manipulate all controls was 4.45 seconds, considerably more than is allowed by the FAA.

During the past ten years there have been 333 jet accidents. Eighty percent of the fatal ground accidents were attributed to the takeoff portion of the flight. All of the aircraft should have been able to stop in time, according to their weight and runway length. The fact that none did leads to the obvious conclusion that the standards for refused takeoffs do not allow sufficient time for the complex decision required.

On November 23, 1964, a TWA Boeing 707 experienced mechanical difficulties and started to abort takeoff well below the critical stopping speed. On paper the plane should have stopped; instead it slammed into construction equipment at the end of the runway and burst into flames. Fifty people died.

Another safe takeoff roll, on paper, happened to Cathay Pac Airways in 1967, at Hong Kong. A Convair 880 experienced severe vibration at a speed of 13 knots below V1. It could not be brought to a stop within the remainder of the runway. All four engines were torn off and the fuselage was cracked in two places. The aircraft was totally destroyed.

In spite of some realistic and tortuous testing, the aircraft manufacturers can never approach the true test of daily airline operations for discovering the weaknesses of any man-made machine. It requires a constant monitoring by all facets of the industry to ensure that the operational life of an aircraft is safe and sound.

At the end of World War II, a new four-engine transport was introduced by Douglas, the DC-6. At the height of its popularity this craft was carrying more passengers than any other transport in the world and was well on its way to being the best money maker of all time. Then disaster struck: a United DC-6 was cruising over Bryce Canyon, Utah. The captain noticed that his fuel tanks were not emptying evenly so he decided to balance the load. He opened the cross-feed and turned on the appropriate fuel-boost pumps. Since this process takes a few minutes he was not concerned with the transfer. He did, however, wait longer than usual to check on the results of his operation. The No. 3 tank had overflowed and was pumping fuel through its only exit, the tank vent. Fuel streamed from the vent and the airflow over the wings fed it directly into the cabin heater air scoop. Now the heater was being fed by gallons of raw fuel. The next time the cabin heater cycled on, it triggered an uncontrollable fire.

The pilot made a desperate attempt to get on the ground before the weakened structure let go. He spotted a field and was descending rapidly when his luck ran out. Less than a mile from the runway, the aircraft blew up, killing fifty-two.

The Civil Aeronautics Administration (predecessor of the FAA) should have been held totally responsible for this disaster. During the design stages of the DC-6, a federal regulation stated: "It shall not be possible for fuel to flow between tanks in quantity sufficient to cause an overflow."

Another part of the regulation read: "Vents and [fuel] drainage shall not terminate at points where the discharge of fuel will constitute a fire hazard."

CAA inspectors had allowed flagrant violation of their regulations. All DC-6s were grounded and the position of the vents moved.

About this time another airliner, the Lockheed Constellation, started to develop bugs. A series of electrical fires occurred, ending in the fatal crash of a TWA training flight near Reading, Pennsylvania. The culprit of these fires turned out to be a through stud, a tiny electrical conductor that allowed power to flow through the pressurized fuselage without destroying the pressure seal. These studs were of "crudely deficient design" (CAB official report). The studs came in such close proximity to the fuselage that arcing was possible. This position was adjacent to a high-pressure hydraulic line thereby increasing the fire hazard. The CAB found evidence of faulty studs in nearly half the Constellations in service, and a conglomeration of unapproved washers that were the catalyst to the fires. Once again the government grounded a faulty airliner. The "Connies'" electrical system was completely revamped.

On August 29, 1948, a Northwest Martin 202 crashed while in the vicinity of a thunderstorm near Winona, Minnesota. The left wing was found a considerable distance from the rest of the wreckage, signifying structural failure.

The CAB examined the wing section and found the telltale clues of metal fatigue. The manufacturer found this hard to believe because the plane had only 1,321 hours in the air. The CAB ordered all of Northwest's fleet inspected immediately. Nearly half of them had cracks similar to the airplane that crashed, and three of them had severe cracks in both wings. All Martins were grounded while the "bug" was repaired.

A badly designed wing flange was the culprit. The CAB stated that the flange was "inducive to high local stress concentration and hence was readily susceptible to fatigue." The CAB inspector who approved the flange had allowed faulty vital structure to be incorporated into the Martin's wing. This mistake took the lives of thirty-seven people.

At least it can be said that the government in those days had the guts to call a spade a spade. It takes an act of Congress to ground an airliner today. There is so much pressure from the manufacturer and the airlines to keep the fleet in the air that, to my knowledge, there has not been a grounding since 1950, and there most certainly should have been.

In December 1957 the Lockheed Company rolled out the pinnacle of their craftmanship. A $2.4 million four-engine stubby-winged turboprop transport called the Electra. In the words of the one-time head of the FAA, Najeeb E. Halaby: "No other aircraft in the history of United States commercial aviation has flown through more misfortune than the Lockheed Electra."

At first there had been only minor operational bugs in the Electra. Vibration was felt in the seats most nearly aligned with the four giant paddle-blade props. This vibration was not detected during the many hundreds of hours of prototype test flights and came to light only during the rigors of on-the-line flying. Here an aircraft is flown by experienced pilots, and by fledglings. Some make the wings quiver when they land it, others grease it on like a feather. Here is the true testing ground for the craft. Will she stay together in tortuous turbulence in varying degrees of temperatures? Will the engines and mounts cope with the experienced hands that gently apply the throttles, or will they resist the rough treatment that a nervous "company man" inflicts on them? Only time would tell.

On September 29, 1959, Braniff Airlines Flight 542 was cruising at an altitude of 15,000 feet, airspeed 275 knots. The radio transmissions were all normal and there was no sign of impending disaster. At 11:07 P.M. a blinding flash was seen in the sky over Buffalo, Texas, and Flight 542 began to break up and fall to the earth. A wing was found in a potato patch a mile away from the rest of the wreckage.

The investigating team did not rule out the possibility of a bomb on board, but the frequency of metal fatigue cracks on the other Electras inclined them to favor this latter theory. After six months of painstaking sifting and examining of wreckage, thc investigators were stumped. They were about to place the tragedy on their "no explanation" list.

Then, on March 17, 1960, at 3:25 P.M., another Electra lost a wing and plunged to its doom near Tell City, Indiana. Northwest's Flight 710 lay buried in a crater 12 feet deep along with the remains of 21 women, 33 men, 8 children, and 1 infant. The total deaths brought by structural failure in flight now reached 97.

If these unsolved accidents had happened in the days of the CAA, the Electra would unquestionably have been grounded. But its successor, the FAA, was being run by a man who evidently never made mistakes, General Pete Quesada. The CAA had certified the plane as being fully airworthy, so he was not about to ground it.

The first FAA order involving the Electra was dispatched three days after its most recent disaster. The Electra speed would be kept below 275 knots. Quick thinking by Captain Robert N. Rockwell, a member of the Northwest ALPA accident investigating team, reminded the FAA that 275 knots was the same speed the Braniff pilot was using when the aircraft began to break up.

Senator Vance Hartke of Indiana called Quesada: "I assume you're going to ground the Electra."

"No, I'm not," Quesada barked.

Angered, Hartke said, "If another Electra crashes, it will be your responsibility."

"I'll accept that responsibility, Senator," Quesada replied.

Pressure to ground the aircraft was coming from all sides, not just Capitol Hill. Edward Slattery, veteran press officer of the CAB Bureau of Safety, championed the cause for grounding.

Quesada evidently had given some thought to Captain Rockwell's statement about the 275-knot speed being too fast and issued a mandatory slower one.

On April 15, Senator Hartke, who by now was engaged in a heated congressional "ground the Electra" crusade, phoned Quesada and with no preliminary chitchat asked, "Is it true the CAB recommended grounding?"

"Yes," shouted Quesada.

"What are you going to do about it?"

"Nothing," said the boss of the FAA.

In Robert J. Serling's book *The Electra Story* there is a passage I would like to quote concerning an interview Quesada had with Serling months after leaving his FAA post.

Quesada: "I never considered the economic effects of grounding on the airlines involved," he said. "Frankly I don't think they could have gone through it without some harmful effects." Guilty conscience?

Quesada goes on, "It's possible, if I had grounded the Electra some people might have been killed in some weary old plane pulled out of mothballs."

That doesn't say much for the enforcement of the FAA regulations which demand that every passenger be afforded equal safety, no matter what age or type the plane may be.

The Electra was flight tested again and millions of dollars were spent to research the wing structure. Millions of people were still allowed to fly on a plane that might at any moment disintegrate in mid-air.

Engineers found that the Achilles heel was in the outboard engine nacells. During certain turbulent flight conditions a vibration would be caused from the outer engines to the wing tips that was greater than the wing could stand. This was known as "whirl mode." Tests both on the ground and in the air went on for over six months. The beefing-up process would cost Lockheed $70 million but it was necessary in order to get the fleet back to cruise speed again.

On January 5, 1961, the FAA approved the modification and lifted the speed restriction. During this giant modification drive the quality control at Lockheed left something to be desired. The FAA fined Lockheed $6,000 for sloppy assembly practice. Eastern Air Lines found the following items in two of the wing tanks of their Electras: twenty small screws, a vacuum cleaner, and a rivet gun. American Airlines had bolts missing from different sections in thirteen of its Electras.

The discrepancies just mentioned were minor in nature compared with the tragic and needless deaths of thirty-seven people aboard a Northwest Electra. This time it was caused by faulty maintenance. After takeoff at Chicago's O'Hare Airport the aircraft rolled over on a wing and crashed. The frantic eight seconds of radio transmission from the pilot to the tower came in a short, high-pitched voice:

"We're in trouble . . . uh . . . all units holding . . . Northwest Electra, I still don't have release . . . right turn in . . . no control . . . no control . . ."

Aileron trouble (devices on each wing that bank a plane) was immediately suspected. Investigators found the right aileron had jammed before impact, locking the plane in a tight turn near the ground. The CAB disclosed that six weeks before the accident, an aileron cable wire was disconnected, and a safety wire which gave the proper tension during the up movement in the aileron was not replaced. Three different shifts of mechanics worked on this section of the plane. The mechanic who had removed this vital wire was discovered, but no one could be found who remembered replacing it. Because of this carelessness thirty-seven people are no longer with us.

The following two accidents vividly portray how economics rule the aircraft industry, even to the point of manslaughter. It is indeed shameful that a lesson learned after one crash can go unheeded for some sixteen years until it once again appears on the same carrier.

Captain George Warner, thirty-five, had been flying as a captain for United for six years. On the morning of June 17, 1948, thirty-nine passengers boarded his Flight 624 from Chicago to LaGuardia.

The weather was perfect and the flight routine. About twenty minutes out from New York, in the vicinity of Allentown, Pennsylvania, the following message was picked up by ATC: "New York! New York! This is an emergency descent."

About 40 miles from Carmel, Pennsylvania, a telephone lineman, Elmer Stahl, saw a large silver plane traveling at a great rate of speed and at a low altitude. Edward Janoka, who had been a rear gunner in the air force, ran out of his house when he heard a "terrific roar." The plane was down to 300 feet now and making aimless turns. It crashed through some power lines and met the ground for the last time at Mt. Carmel, Pennsylvania.

Mrs. DeVito, the wife of one of the deceased passengers, hired a well-known lawyer, Harry A. Gair, to see what could be done about getting some restitution from the people who had killed her husband. Mr. Gair brought suit against both United and Douglas.

On February 5, 1951, a jury was impaneled before the Honorable Clarence G. Galston, in the U.S. District Court for the Eastern District of New York. United was represented by William J. Junkerman, himself a pilot and veteran of many trials involving aircraft disasters. Douglas had retained Mr. Lasher B. Gallagher.

The attorneys for Douglas fought vehemently against using the last recorded radio message from Flight 624. They said that in its present state the tape was not clear and they wanted to prevent experts from eliminating the high background noise on the tape. Finally, after days of discussion, the tape was allowed as evidence. The fact that the pilot had radioed that he was making an emergency descent added to the case against Douglas. Perhaps this plane was not as the manufacturer claimed.

The tape was played for the court and it left no question as to the emergency. In a voice choked and strangling the pilot spoke. "624 . . . 624 . . . Fire! Discharging . . . forward baggage pit . . . 624 coming down! I . . . ah . . . ah . . . coming down!"

The beginning of the mystery was traced back to the fall of 1947, when the DC-6 had two uncontrollable fires in flight. During January and February of 1948, tests were conducted by Douglas to improve the fire-fighting capabilities of the plane. The basic ingredient in the extinguishing agent was carbon dioxide. This efficient chemical prevented the combustion of oxygen.

These tests were only in their infancy when the fearful fact was discovered that the fumes from carbon dioxide posed a lethal danger to the crews. Test pilots were reporting that after discharging the bottles into the lower baggage compartments, the deadly fumes were seeping through the cracks in the floor. The amount of this gas on the flight deck was so great that the incapacitation did not take long. Captain L. Patent, a Douglas test pilot, stated that the pilots had to go on oxygen before discharging the fire extinguishers.

The Douglas Company contacted Dr. Clayton White, of the famous Lovelace Clinic in New Mexico. His tests proved conclusively that the discharge of the carbon dioxide, a colorless, odorless, tasteless gas, in a confined space may produce collapse and death. (Most of our modern jets still use carbon dioxide to combat fires.) The tests pointed clearly to the disaster to come *six months later*.

Under examination by lawyer Gair, Douglas' chief pilot was asked who had written on the vital report from Dr. White to Mr. Douglas: "Do not discuss this with the CAB." He readily admitted that he had. When asked if he had discussed the reports with the CAB he said he had not.

On May 31, 1948, United conducted their own tests on the DC-6 and gave it a clean bill of health. The baggage compartments leaked a little, but not very much, so on June 3, 1948, the plane was recertified and the stage was set for the disaster. Two weeks later, the fatal crash took place.

Seven days before this, Allen W. Dallas of the engineers' division of the Air Transport Association sent telegrams to all operators of the DC-6 and the Constellation: "TWA Constellation, forced landing at Chillicothe, revealed serious oxygen deficiency in cockpit when CO_2 bottles were pulled in cargo compartment. Under all circumstances oxygen should be used before releasing CO_2. Similar situation may exist on DC-6 aircraft."

Mr. Douglas's reaction to this was to protect the honor of his plane. He demanded that a follow-up wire be sent to all operators stating that Douglas had run extensive tests on their aircraft and had proved it airworthy. He demanded that any mention of the DC-6 be stricken from the telegram.

On June 16, the *day* before the accident, United's conscience began to bother them. A bulletin was prepared to warn its pilots: "Use oxygen masks before discharging CO_2." Any message as vital as this one should have been teletyped to every pilot base and posted in a conspicuous place. Instead it was sent to the print shop to be mailed to each individual crew member.

If the Douglas people had disclosed their findings to the industry, and if United had acted faster on a crucial matter, there would not have been any need for a lawsuit. Both Douglas and United were to blame, and Douglas had to pay $140,000 to the DeVito family.

It is one thing to make a mistake, and at times there may be extenuating circumstances, but to disregard lifesaving procedures is as unforgiving as flight itself. Especially when the well-being of millions of air travelers is entrusted to you.

On July 9, 1964, a United Air Lines Viscount crashed near Pariottsville, Tennessee. Thirty-four passengers and a crew of four perished. Cause: uncontrollable fire in flight which resulted in a loss of control of the aircraft.

The investigating team once again ran the gamut concerning the fire-extinguishing agent, carbon dioxide. The most serious discrepancy noted during the tests concerned leakage of CO_2 into the cockpit when extinguishers were discharged. The bottles containing the suppressant were located behind the copilot's seat. When activated, they were supposed to discharge into the lower baggage compartment. Instead, after fifteen separate tests a lethal amount of CO_2 was admitted to the flight deck. Many of the seals that govern the flow of CO_2 were found installed improperly, thereby allowing the gas to flow in the wrong direction.

The FAA was given copies of all the findings that the CAB had uncovered during their flight and ground tests. The FAA published a proposed aircraft directive concerning the extinguishing system of the Viscount. This proposal appeared in the *Federal Register* of June 29, 1965. All rebuttals from interested parties concerning the proposed rule would be considered up to July 27, 1965. This meant that the operators of the Viscount had a month to decide if they would like to have the proposed rule changed or modified. While management considered the economic penalties involved, a potentially dangerous situation was allowed to continue unchecked.

By October 3, 1965, the FAA finally decided that something should really be done about the faulty extinguishers. They published a directive which stated: "Compliance required at first airplane overhaul or within six months after the effective date of the Airworthiness Directive, whichever occurs first, unless already accomplished."

At least the requirement concerning the pilots donning their oxygen masks was emphasized. But the time lag allowed by the FAA, over six months for a vital fix, is inexcusable.

The federal air regulations concerning a class D compartment plainly states that compartment must be designed so that a fire will be contained within the confines of the compartment and that it must be constructed of fire-resistant material. Practically all modern jets utilize the class D compartment; it is where the cargo and luggage are stowed below the passenger floor.

In essence these compartments consist of a thin sealed baggage bin within the pressurized belly of the craft. It has been proved that at least 10 percent oxygen is required to sustain combustion. Being nearly airtight, these compartments admit a small supply of oxygen from the pressurized envelope—enough to keep animals alive if they are shipped by air freight, but supposedly not enough to sustain a fire.

When the aircraft is new, these compartments live up to their requirements, but when cargo doors have been slammed repeatedly, and skis and poles have been thrown into them, the once-tight compartments are a thing of the past, and a new look resembling a sieve dominates. These compartments give no fire warning in the cockpit, and also contain no extinguisher. The FAA says a sustained fire is impossible so there is no need for either of the aforementioned safety devices. Or is there?

Next time you're on a 707 and the floor seems excessively warm, it is comforting to know that the blazing fire that might be down below is unknown to the pilot, and his only course of action is to land immediately before he and you are just another statistic.

Captain Arthur Moen, forty-seven, First Officer Johannes D. Markstein, thirty-eight, and Flight Engineer James R. Skellenger, thirty-one, have all made their final flight for Pan American Airways. On December 26, 1968, their 707 crashed on takeoff at Anchorage, Alaska. A $10 million machine and a crew that no money can replace lay in a fiery heap at the end of runway 23.

A large part of the blame can be attributed to Pan Am, and more directly to one of its engineers. Investigators discovered that the all-cargo flight attempted to take off with the flaps in the "up" position. All large aircraft are required to carry a given amount of flaps for takeoff. This aerodynamic lift-and-drag device enables a plane to take off in a much shorter distance.

The cockpit voice recorder was recovered and played. There was no audible sound of a flap takeoff warning horn. This horn is a backup for the check list. If it blows when the throttles are advanced it immediately signifies to the pilot that one of three items is not in its proper place: flaps, spoilers (sections on top of the wing that extend to reduce speed on landing), or tail trim setting. Why didn't the horn blow?

During the early part of the investigation, it was determined that Boeing Company issued a service bulletin, #2384, on January 31, 1967, twelve months before the fatal accident. This bulletin was sent to all operators of 707 equipment. It stated that in cold-weather operation the horn would not blow unless the actuating stops of the throttles were reduced from 42 to 25 degrees of thrust-lever advancement. A similar service bulletin was also transmitted to cover the 727 series aircraft. (During cold weather the air is more dense and higher power setting will be obtained with less movement of the throttle. In this case the throttles weren't advanced far enough to actuate the horn but still the required thrust was obtained.)

The service bulletin issued by Boeing carried the word "recommended" rather than stronger advice. Boeing obviously felt that the various airlines would comply with such an important request. The FAA should have demanded it.

One of Pan Am's engineers received the service bulletin from Boeing and questioned other interested people in his company, i.e., maintenance and flight operations. Some of the factors considered during the review were the frequency of exposure their craft would be subjected to and the relationship of safety to this bulletin. If a determination was made to adhere to the directive and the total cost was under $500 (as it was in this instance), a modification request would be prepared.

However, an engineer decided that the bulletin was not applicable to Pan American and no further action was taken.

In response to the safety board's inquiry concerning the reasons for the nonimplementation of the service bulletin, Pan Am cited the following reasons.

1. There was no aircraft crash on record relating to the lack of this modification.
2. The FAA did not demand that the bulletin be complied with.
3. Other air carriers also elécted not to comply with the bulletin.

Pan Am should have been ashamed of such lame excuses.

A playback of the tape showed that Captain Moen was indeed following company procedures when he started his taxi to the runway. The F/O was running the check list and placed the flaps in takeoff notch, but Captain Moen returned them to the full "up" position. The company Operations manual states: "The wing flaps should be left in the full up position until lined up for takeoff. This would reduce the chance of snow and ice being blown onto the flap mechanism."

Positioning the flaps for takeoff appears only on the taxi position of the check list. There is no mention on the pre-takeoff section of the list to remind the pilot to lower the flaps before it is too late. FARs plainly state that a pilot must not rely on his memory alone to perform various functions. A check list must be used. Here is a conflict between FARs and an adequate check list.

The check list used by Pan Am and many other carriers leaves a lot to be desired. There are in use today three distinct types of lists: (1) a plastic card; (2) a scroll; and (3) the most expensive and best, the mechanical list. This check off list embodies a lighted background and a series of tabs that can slide over the telltale light when the required item is completed. If, as in the case of Pan Am, an item is to be skipped, such as the flaps, the tab will remain in its stowed position signifying at a glance the parts of the list that have not been completed.

On May 28, 1969, five months after the accident, the FAA

finally got around to issuing an airworthiness directive that made it mandatory to comply with the horn setting on all 707 and 727 series aircraft.

Once again Alaska is the scene of needless disaster. On December 2, 1968, thirty-nine people perished as the wing of a Wien Consolidated Airliner Fairchild 27 fell off in flight. The NTSB determined that the failure was caused by severe turbulence that tore off a wing already weakened by preexisting cracks. The board recommended to the FAA that all Fairchilds with over 5,000 hours in the air should be checked for fatigue. This recommendation was followed; of sixty-seven aircraft inspected, a total of thirteen cracks were discovered in eight of them.

A thorough review of the company radiographs (similar to an X ray) revealed that cracks were present in the wing structure in October 1967, fourteen months prior to the accident. The October cracks were small, but as they went unchecked their size increased. In April 1968 the number of cracks had multiplied from two to seven. The last radiographs taken in October 1968 showed nine more cracks at four different locations in the wing. A review of the maintenance logs showed no written record of the existence of any of these cracks.

The aircraft was not in compliance with the existing airworthiness directives at the time of takeoff in its last flight. And it had not been airworthy for some time before it crashed. The special inspection conducted by the FAA indicated that similar cracks existed in aircraft operated by other airlines and that these had also not been noted in the aircraft records.

Other interesting facts concerning the power plants were discovered by the board. When an aircraft engine is shut down the propeller will windmill in the airstream and cause excessive drag. This could grossly affect the flight. If an engine failed to develop sufficient power, the blades of the propeller would automatically knife through the air in a feathered position.

This was a relatively trouble free system. But the propellers on the Rolls Royce Dart engine that powers the Fairchild and Viscount aircraft will also feather for another cause. If the airplane should be placed in a sudden updraft for no longer than two seconds, an engine could shut itself down and feather automatically. This is just the way these particular power plants operate, and it most certainly should never have been certified as airworthy while it contained a booby trap such as this.

On a twin-engine plane, if one engine feathers the other engine will be prevented from following suit by a blocking device. However, there are two four-engine aircraft—the DeHavilland Argosy and the Vickers Viscount—that are subject to even more horrendous difficulties than the twins. If they should be buffeted about in turbulence and are forced by an updraft into the negative 0.1 g for two seconds, all four engines are likely to shut down.

The Rolls Royce people are cognizant of this bad characteristic and in March 1965 published operating instructions as follows: "Negative 'g' maneuvers should be avoided. If, however, sustained negative 'g' flight is encountered, close both throttles to idle until normal flight is resumed."

That might be a ticklish trick, especially if the power is essential to climb over an obstacle, or immediately after take-off.

The Grumman Gulfstream, the Convair 600, and the Convair 640 are also powered by the Dart engines. A total of five-hundred craft are affected in the United States alone. As long as the "g" load is so important, it would be nice if the pilots were afforded the opportunity to check their "g" load with a gravity instrument. These gauges are extremely cheap and would undoubtedly be of some benefit to the pilot. No U.S. carrier has these instruments on any of their craft. The only people that are afforded the luxury of a G meter are the military.

At the time of the December 1968 accident in Alaska it is believed that one of the engines feathered as well as losing a wing. There are other case histories of these Dart propellers stopping during scheduled airline flights. The NTSB fired off a letter to D. D. Thomas, who was then head of the FAA, alerting his office to the grave dangers inherent in the engines. His ludicrous reply concerning an aircraft that has no business in the sky is as follows:

> This problem has been under evaluation by the FAA since receipt of information from the Navy on September 30, 1968. As a result of these evaluations we have requested the manufacturers of the affected aircraft to prepare and issue appropriate airplane flight manual revisions which will warn the pilot of the possibility of propeller autofeathering in negative "g" encounters.

Evidently Mr. Thomas did not carefully read a letter addressed to him on December 30, 1968, from Mr. Joseph J. O'Connell, chairman of the NTSB, wherein the special instructions concerning Rolls Royce Dart engines were clearly stated. These engine revisions were sent out to users in March 1965, yet it took a navy plane's mishap in September 1968 and a letter from the chairman of the board in December 1968 to get some action. The Fairchild is still flying and to this day nothing has been done about the auto-feathering.

The Fairchild has many other glaring inadequacies. It is probably the only airliner in service today that fails to alert the pilot properly to the fact that his tail heater may be on fire. The usual method of warning a pilot that part of his plane may be on fire is a red warning light and a bell. The Fairchild does have warning lights for both engines and the cabin heater in the tail, but there is no extinguisher for the tail, only for the engines. A warning bell has proved to be the best means of denoting trouble. When the sun is shining into the cockpit, it is extremely difficult to see a tiny light. The crew could conceivably be too busy even to notice the light. A bell can be heard no matter what the workload.

The total cost for one airline for this modification would be $748.05. Yet management would rather lose a plane than spend the money.

The total cost of a switch guard is under $5 and the necessary installation time less than ten minutes, and yet this all-important modification was left until it was too late.

On July 6, 1969, an Air South Beechcraft B-99 crashed

near Monroe, Georgia; twelve passengers and both pilots were killed. The NTSB determined that the probable cause of the accident was an unwanted change in longitudinal trim which caused the nose to drop rapidly in a dive that was beyond the physical capabilities of the pilot to overcome.

Many operators using this type of aircraft had experienced trim system malfunctions. In all incidents a violent pitch-up was experienced.

On July 8, 1966, Beech Aircraft Company applied for a type certificate for their 99 model. This plane was approved for scheduled passenger service by the FAA under their delegated option authority. This plan allows the aircraft manufacturer to certificate his own product. The FAA did not scrutinize all of the certification data given it by the company, only such areas as they felt necessary. They did not witness all the flight tests. The FAA is required, even under the lax testing methods, to participate in testing when a previously uncertified design feature is introduced, such as the trimmable stabilizer in the Beech 99. However, the FAA failed to witness any of the certification of this system, the system that caused the crash.

On November 24, 1971, another Beech 99 ran into trouble when one of its propeller blades broke. The out-of-balance rotation of the prop caused the right engine to be torn from its mount. Fortunately, a safe landing was made and there were no injuries. Metal-fatigue cracks were found in the blade that were present before the accident.

Beechcraft is at present up to its neck in lawsuits over allegedly faulty fuel tanks. Four models are supposed to contain gas tanks that will not properly feed fuel to the engines when the aircraft is placed in certain attitudes. Beech say that "several thousand" of this type have been sold. A class-action suit filed against the company alleges that over fourteen-thousand planes have the defective fuel system. Over nineteen people have been killed in crashes of Beech planes since 1964. The first known suit confirming the allegedly faulty system was settled out of court for $297,000 according to the lawyer retained by the estate of the pilot who was killed. Beech people said they had no information on the settlement.

Another case, filed in June 1971 in Santa Ana, California, Superior Court, culminated in a $21.7 million judgment against Beech. It is believed to be one of the highest ever awarded in such a case.

In January 1967 the NTSB recommended that the FAA inspect the fuel system on some Beech models. Two years later, and after more deaths, FAA test pilot Stewart Present concluded that "a potentially hazardous situation exists in this aircraft" (referring to the Beach *Baron*).

Another case of FAA laxness is evident in the following crash. On January 8, 1970, a DeHaviland Dove 104 airliner used by TAG Airlines crashed in Lake Erie. All on board perished. The NTSB found that the right wing of the craft had sheared off in flight. The examination of the wreckage left no doubt that the chrome-plated right-wing lower root joint had numerous fatigue cracks *prior to the accident*.

The manufacturers of the Dove, a British Company, were long aware of the faults that the chrome plating of this fitting would cause. In 1961 they recommended that any chrome fitting be changed the next time the wing was removed, or before 10,000 hours, whichever came first. This was mandatory for Doves registered in the United Kingdom.

Based on this recommendation the FAA also issued an airworthiness directive in September 1961. However, they only adopted half of the British recommendation. The FAA said that the chrome fittings should be replaced before ten thousand hours, but no mention was made of a change when the wing was removed for routine maintenance. When the wings of the Dove aircraft involved in the Lake Erie accident were removed for maintenance, the faulty attach fittings were not replaced. The total time on the wings was approximately 9,384 hours, 616 hours short of the requirement set by the FAA. If the FAA had taken the manufacturer's safer approach the crash would never have happened.

On May 6, 1971, at Coolidge, Arizona, another Dove

crashed and all on board were killed. This disaster is still under investigation, but there was no question concerning the same fitting that was at fault on the previous Dove crash. A metallurgist from the NTSB made this statement concerning the latest disaster: "The right wing lower attach fitting was found in a broken condition and the fracture stemmed from a pre-existing fatigue crack."

Captain Charles E. Bullock, forty-three, a most competent senior pilot with Mohawk Airlines, flew his last flight on June 23, 1967, Flight 40 from Elmira, New York, to Washington, D.C., with a BAC-111, an aircraft that is the British counterpart of the DC-9. Captain Bullock was ably assisted by First Officer Troy Rudesill, thirty-three. His logbook total showed 4,814 hours, only a third of the time "Chuck" Bullock had in the air.

The takeoff was normal and everything was routine until approximately five minutes after departure. At 6,000 feet things started to happen.

"Ah, let's see pull back on your speed," said Captain Bullock. His first officer was flying.

"Wait a minute, I'm doing it . . . heh!"

"There's something screwy here."

"I know it."

Chuck diverted his attention from the predicament at hand and contacted the air traffic controller: "Mohawk 40 is having a little control problem here. Will ah, will advise you, ah we may have to declare an emergency."

Probably the biggest worry pilots have concerning an in-flight emergency is fire, with control problems running a close second. Unaware of a raging fire in the tail section, Captain Bullock was in the grips of both emergencies at the same time. As the gravity of the situation became more intense, Chuck took over the controls.

"I got it. I got it," Captain Bullock said.

"Righto."

"Tell 'em we're unable to maintain 6,000, going back to Elmira."

Captain Bullock's composure was commendable throughout. Here is where the years of experience really come to the surface. Chuck said anxiously: "We've lost all control! We don't have anything.

"We're in manual now." [The normal hydraulic boost control was inoperative.]

"Yeah, but I can't do anything."

"Well—okay."

"You turned it off!? Put it back in second system," Captain Bullock said. "Let's go up for a minute. Here we go—easy now, easy."

The ill-fated craft was pitching up and down in ever-increasing oscillations. Now it was down to 4,000 feet.

"Pull back!" Chuck yelled in an emotion-filled voice. "Pull back! Keep working, we're making it. Pull back—straight now. Climb!"

Now the crippled whale started to climb again to 5,500 feet.

"Cut the gun [retard the throttles]. Cut the gun, we're in now," Bullock said.

"Okay."

"Oooh-wee! I don't like that! What have we done to that damn tail surface, ya have any idea?"

"We have lost both [control] systems."

The airliner was in its final dive and rapidly reaching for the ground. Now came the stark realization that death was not too far off.

The voice recorder picked up the final words of a most gallant crew: "I can't keep this—all right. I'm gonna use both hands now."

"Okay."

"Pull 'er back, pull 'er—both hands back. Both hands back! PULL BACK. I'VE GONE OUT OF CONTROL."

The aircraft began to shed parts in flight due to the terrific overspeed. It contacted the ground at a 40-degree angle and cut a swath through the trees 1,300 feet long and 420 feet wide. Blossburg, Pennsylvania, became the final resting place for Mohawk's Flight 40: thirty passengers and a crew of four.

After months of research and testing the NTSB found the probable cause and remedy. A faulty no-return valve was the culprit. The auxiliary power unit (APU) contained in the tail supplies the aircraft with electricity, pneumatics, and air conditioning while on the ground with the engines shut down. It could also be used in flight to perform the same functions as it does on the ground. There is absolutely no need to run this high temperature turbine in flight since the aircraft engines supply all the systems needed.

At the time of the accident the company procedure was to run the APU until after takeoff and shut it down on climb-out. In the case of the accident, a valve failed to work and high-temperature air from the unit flowed in reverse through the faulty valve into the APU and exited into the airframe plenum chamber. This intense heat burned through the support section for the rudder and in so doing a hydraulic flex line was singed and sprayed hydraulic fluid (which is not supposed to burn) over the APU. A death-dealing cycle was established which burned part of the tail off in flight and rendered the control system inoperative.

Since the discovery of this faulty valve in the APU, a worldwide campaign has been undertaken to check the no-return valves in some eighty BAC aircraft. Sixty percent of the valves checked were faulty, but still serviceable; 20 percent were completely unairworthy.

On the basis of the recorded cockpit conversation, the NTSB determined that no mention of fire was made by the ill-fated crew. There was no fire detector to warn the crew of a fire in the tail. *There still isn't!* The big fix for this catastrophe consisted of removing the soundproofing material enclosing the exterior of the APU with steel and replacing the faulty valves. No mention was made of putting a fire detector in the vicinity of the APU.

Since this accident most pilots have been hesitant to use the APU in the air, for which they certainly can't be blamed. About a year after the accident a captain was route-checked by his chief pilot. He failed to use the company procedure which included running the APU after takeoff. Many pilots used the far safer procedure of turning it off just after starting the engines. However, the chief pilot insisted that this captain be sent to their head offices for a day's ground school on the use of the APU. Here he would be brainwashed into how safe it was since their modifications.

The engineers and check pilots were all on hand to persuade this pilot to use the APU in the air. A few months after the harassment session, Mohawk Airlines again had trouble with the APU. A BAC-111 was parked at the gate in Minneapolis, Minnesota, with the APU running. Were it not for an alert ramp agent, the empty plane could have burned to the ground. There was a failure once again in a component of the APU turbine and it caused a fire that nearly burned the bottom of the plane through.

Now the same chief pilot had to eat humble pie. He ordered all of his pilots to restrict the APU to ground use only. Slowly the Mohawk fleet is being modified again so that the APU can be used in flight—which is not needed except in an emergency. They are installing an automatic thermal shutdown so that the APU will stop if the temperature exceeds a given amount. That is a fine idea, except there is *still* no fire detector in the vicinity of this turbine to warn the pilot of a fire that may be burning long after the APU has shut itself down.

If Captain Bullock had been afforded the luxury of a fire detector in his plane he would probably be alive today. It is believed that the fire was probably raging inside the tail before he ever applied the power for takeoff. It's an amazing thing in aviation how the impossible malfunctions come to light many times before the FAA takes any suitable action.

On January 18, 1969, United's Flight 266, a Boeing 727, left Los Angeles International terminal and received taxi clearance to runway 24. An extended area of fog, rain, and low clouds prevailed along most of the California coastline. The official airport weather was rain and fog, 700 feet scattered clouds, measured 1,000 broken, and a visibility of 3 miles.

Captain Leonard Leverson, forty-nine, had recently completed upgrading training on the DC-8 and had not flown a 727 for forty-seven days. (FAA allows a pilot to go as long as three months without flying before becoming unqualified.) His first officer, Walter R. Schlemer, had logged nearly 2,000 hours in the tri-jet, and Engineer Ostrander, who was relatively new at the game, had barely 500 hours, only 40 of them in the Boeing 727.

Captain Leverson had acknowledged his permission to taxi onto the active runway. At 6:16 P.M. the tower issued the following instructions: "United 266 cleared for takeoff."

F/O Schlemer answered, "United 266 rolling."

The few remaining check list items were completed.

"Engine start switches."

"3 on."

"Antiskid."

"On. Yeah, that's good."

"Oil cooler coming ground off."

"They're stabilized."

Captain Leverson said, "Take off thrust."

F/O Schlemer replied, "Set looks good."

As the craft gathered speed down the rain-soaked runway the jargon familiar to all pilots was heard as the F/O called off vital speed information: "100, 110, 120, VR, V2."

Captain Leverson gave the thumbs-up signal and called, "Gear up."

The tower operator's voice came over the cockpit speakers: "United 266, contact departure control."

"United 266 on departure."

"United 266, Los Angeles departure radar contact, turn right heading 270 degrees, report leaving 3,000."

"270 degrees wilco."

The efficient cockpit crew were startled by flashing red lights and the spine tingling CLANG! CLANG! CLANG! of the emergency system.

"What the hell was that?" asked the Captain.

"Number one fire warning," Walter shouted.

"OK, let's take care of the warning."

"That puts us on one generator," said Walter.

As Leonard answered his first officer he was already starting a turnback to the field: "Yeah, yeah, watch the electrical loading."

The flight was now over the Pacific Ocean and flying on instruments at night and in the clouds. Walter was again working the radio: "Ah—departure, United 266."

"United 266 go ahead."

"Ah—we've had a fire warning on number-one engine, we shut down, we'd like to come back."

"United 266, Roger. What is your present altitude? United 266, maintain 3,000 and say your altitude. . . . United 266, say your altitude."

There was no time for the flight crew to bother with radio contacts. Far more important things were happening in the cockpit. The engineer's panel resembled a giant pinball machine. Warning lights were flashing rapidly, and there was little time to cope with them.

"We're going to get screwed up," said the engineer. "I don't know what's going on." All cockpit instruments stopped functioning. All that remained were the rate instruments, vertical speed, altimeter and air speed. There were no backup instruments. With the existing flight conditions it would be impossible to keep a plane right side up.

First Officer Schlemer, desperately trying to aid his captain, shouted, "KEEP IT GOING UP, ARNIE. YOU'RE A THOUSAND FEET . . . PULL IT UP!"

The departure controller continued his frantic calls: "United 266, if you hear us squawk 0200 or 0400. . . . United 266, if you hear turn right heading 260 degrees."

The air traffic controller knew that the calls were in vain but he could hope. As the sweep of the radar antenna on his scope showed, one of his targets was missing. United 266 had disappeared into the ocean 11 miles off shore. Thirty-eight more deaths added to the list of company and FAA carelessness.

United was experiencing more than its share of engine shutdowns. From 1966 to 1969 over seventy engines were shut down in flight. Only ten proved to be false-alarm warnings. The Boeing Company issued a service bulletin, #25-15, in May 1968. This directive made it optional to replace a senser in the engines so that an increase of 50° F. would be necessary before a warning would actuate the fire bell in the cockpit. This was not done on the affected engine.

In January 1969 the No. 3 generator control panel was removed and replaced with a panel that had been proved defective on eight different aircraft for varying reasons. But this was the one on board the night of the crash. This generator system had a long history of "bugs." In September 1966 a Westinghouse service bulletin recommended replacement of a silicone rectifier for their generators which supplied electricity to this aircraft. This was not done. The No. 3 generator was kicking up such a fuss that maintenance reverted to their old standby, the MEL (the minimum equipment list which specifies the number of components that can be malfunctioning on a flight). So the generator was rendered inoperative. The 727 had its blessings from the FAA to fly on two generators, which it did for a total of three days. The craft had 41 flying hours on it during that period but it was not repaired at any of the stations because of the exigencies of available aircraft and flight scheduling.

When Marvin Whitlock, senior vice president, Operations, for United, was asked his opinion of this matter, he said, "I'm not proud of the 41 hours, but I don't feel that way because of safety."

During the months of June and July of 1969, United Air Lines experienced three more complete generator failures, but they were unable to duplicate each incident. The cause of the failures is still under investigation.

The board made the following conclusion: the entire electrical system was placed on No. 2 generator and it could not stand the load. Yet they still maintain that a two-generator operation is safe. ALPA recommends that "All pilots not accept aircraft with an inoperative generator, as it is considered by the association as unsafe."

The remains of No. 1 engine were thoroughly examined and it was determined that the fire extinguisher warning that the crew received was false, and that no traces of fire were found. If the Westinghouse bulletin had been heeded and an element of higher rating installed, the false warning might never have come on and the accident could have been averted.

There are two vital switches within three inches of each other on the flight engineer's panel: the galley power, which is a tremendous drain on the electrical system, and the battery switch, which controls the essential power to the vital instruments when the generators are inoperative. Since these two switches look identical in a dimly lit cockpit, during a confusing episode it would be extremely easy to turn the wrong switch off, which is one possibility in this accident. As the galley power switch must be turned off to reduce the electrical load during an emergency condition, it would seem only proper that a guard be positioned over the battery switch.

Months prior to the accident, a United pilot wrote a letter to his chief pilot suggesting that a guard be placed over the battery switch before it caused an accident. The answer was to the effect that his was the thirteenth letter to date with the same idea in mind. United felt there was no need for such a guard although his efforts were appreciated. No action was taken on the matter.

On January 31, 1969, the FAA issued an airworthiness directive requiring all B-727 operators to provide a means to prevent inadvertent operation of the battery switch on those aircraft in which the battery control is located within ten inches of the galley power switch.

Battery power can be a vital commodity as the preceding accident proves. Yet the majority of today's jets have this switch crowded on both sides with similar ones; it looks like a row of tin soldiers. They are not guarded, just lying in wait to trigger another accident.

In spite of the warnings given by ALPA about dispatching an aircraft with one of the generators inoperative, and obviously disregarding the crash of one of their planes, United Air

Lines has made no changes in their minimum equipment list. As of January 1971 their MEL still allows a two-generator operation. And that's two years after the lesson should have been learned.

The 747, the largest and most sophisticated airliner ever built, is not without its troubles. What follows are a few discrepancies that are making life miserable for one of the largest carriers. The 747 is guided in all realms of flight by the most complex of computers. These computers are the brains behind the flight instruments displayed in the cockpit. They cost over $100,000 apiece, but what they tell the pilot is another story. One computer may tell him he is diving, another may tell him he is climbing, yet another shows level flight. This leaves the pilot with an eeny-meeny-miny-mo situation.

The four Pratt and Whitney engines that propel this giant are the most powerful ever designed for a transport. They develop 47,000 pounds of thrust as compared to 18,000 pounds for the recently dethroned queen, the 707. The cost per engine is well over a million dollars, but even at that price it is good that they are still under guarantee.

Many of the operators who are flying the 747 have experienced more engine shutdowns during the short while it has been in use than they did during years of service on other aircraft. The main problem is still as it was on the inaugural flight: overheating. One engine got so hot that it shot a sheet of solid flames 15 feet long behind it. Before it was shut down it burned the aft part of the engine and also part of the flaps and ailerons. All three surfaces were so badly damaged they had to be changed.

During landing, when the engines are reversed to slow the plane down, a very serious overheat problem is generated. So much so that if the reversing operation is not canceled at 80 knots, the engine will have to be shut down. There are also engine-bleed problems that have caused compressor stalls during landings. Oil consumption is considerably more than predicted.

In August 1970 an Air France 747 had just taken off from

Montreal on the final leg of its flight to Paris. At an altitude of 5,600 feet a loud explosion was heard. It was the No. 3 engine blowing up. The fire that followed was quickly extinguished by the crew, and an uneventful landing was made at Kennedy.

A month later another 747 en route to Kennedy experienced engine problems, this time far more damaging. American Airlines Flight 14 lifted off from San Francisco International Airport and no sooner were the wheels neatly tucked away when it happened. The No. 1 engine exploded. All electrical cables, fluid lines, and pneumatic ducts located in the engine pod were severed and an intense fire ensued. Both fire extinguishers were shot into the engine with little effect. The blazing fire was seen by all on board as its flames licked at the wings. The fire lasted for three minutes after the last extinguisher was fired. The shrapnel-like pieces of metal from the No. 1 engine damaged the wing and flap area intensively. Fuel was leaking from the wings. It was a miracle that the whole plane didn't blow up.

Shortly after takeoff from Kansas City the left inboard engine of a TWA 747 quit. Seconds later, the right outboard engine burst into flames. Both engines were shut down, and after thirteen hair-raising minutes an uneventful landing was made and no one was hurt.

The 747 flaps were designed to extend to a maximum of 30 degrees; however, because of the complications in the mechanism, all of the carriers are not going past the 25-degree notch. If a pilot should select 30 degrees and he puts the handle in this notch, it requires a thorough flap inspection (which takes hours) before the plane can continue in service. Because of the unusable last 5 degrees of flaps, a 5- to 7-knot higher landing speed must be used. This in itself can cause a potentially dangerous situation since the stopping distance on a marginal runway will be lengthened.

The intercommunications between the cockpit and the cabin are irregular, to say the least. During an emergency the lifesaving instructions from the crew may never be heard. On the lighter side, the sound for the movie being shown in one section of the cabin may be heard by a movie audience in another section.

The APU supplies the essential electricity and air-conditioning while the plane is on the ground. At times in the air the APU door has opened, causing a severe vibration and necessitating the dumping of fuel and a return to the airport.

The FAA sends a bulletin to all interested carriers concerning any problems or emergencies that have occurred on aircraft similar to the ones they might have in their fleet. The ATA also publishes a bulletin similar in nature to that of the FAA. However, both of these organizations make sure that the essential exchange of information is kept secret from the airline pilots. The only offices to receive this vital material are engineering and maintenance. There are airliners flying today that are designed contrary to safety standards. Yet they still license them.

FAA certification of an aircraft is no panacea against bugs! This has been proved numerous times and undoubtedly will continue. I firmly believe that a more extensive testing period should be made mandatory.

The French Caravelle jet was used in cargo service in Europe for two years before it carried passengers. While this is a very expensive proposition, the traveling public would be the immediate benefactors. It is true that many design bugs never appear until the airplane has been in service for years. But, by the same token, the public deserves the added protection.

On July 5, 1970, Air Canada flight 621 was approaching Toronto Airport for a scheduled en route stop. The weather was perfect for flying, and as the giant DC-8 was being vectored toward final approach, the crew was busy completing the last part of the pre-lading checklist.

FIRST OFFICER ROWLAN *Nice day.*

CAPTAIN PETE HAMILTON: *Beautiful.*

TORONTO TOWER: *Air Canada 621 is cleared for a back course 32 turn right to 200 degrees.*

ROWLAND: *Roger* (Then to Pete): *Check 3 green, 4 pressures, spoilers on the flare.*

The final item on the checklist signified that three wheels were in position and locked down and that the spoilers, devices on the wings to spoil the lift and help stop the aircraft when on the ground, would be armed on touchdown.

HAMILTON: Thank you.

SECOND OFFICER HILL: *Spoilers to go and the board is clear.*

HAMILTON: *Thanks.*

The descent angle of the DC-8 was being slowly eased up to allow for the flare and landing. When Captain Hamilton gave the OK, it was the signal for the First Officer to arm the spoilers. Instead, however, he pulled the handle aft and actuated them. The altitude was a scant 60 feet, not enough height to stop the rapid descent rate before contacting the ground. The alarmed Captain Hamilton rammed all four throttles full forward in an effort to stop the dangerous sink rate, but it was too late. The craft hit the runway with such terrific force that No. 4 engine and pod were torn loose from the wing and immediately burst into flame.

ROWLAND (in disgust): *Sorry, Pete.*

TORONTO TOWER: *Air Canada 621 check on overshoot. You can contact departure on 119.9 or do you wish to come in on 5R?*

HAMILTON: *I think we will go around again. I think we'll go straight around.* (To his First Officer) *Get the gear up please.*

ROWLAND: *Toronto Approach, Air Canada is overshooting on 32. We would like to circle back for another attempt at 32.*

DEPARTURE CONTROL: *Runway 32 is closed. Debris on runway. Five is probably the best. The surface wind is northwest about 10 to 15. Turn right 070, 3,000 feet.*

By this time the crew had taken stock of their perilous situation. Amidst the flashing of warning lights and bells, they were fully aware of the danger they were in.

HAMILTON: *We've lost No. 4 engine.* ROWLAND:

ROWLAND: *Have we?*

HILL: *No. 4 and No. 3 is jammed, too.*

ROWLAND: *What was that? What happened then, Pete? Look, we've got a . . .* (Sound of loud explosion aboard the plane.) *Pete, sorry.* (Sound of explosion even louder than before.)

HAMILTON: *All right.*

DEPARTURE CONTROL: *The status of your aircraft, please.* (Sound of metal tearing apart.)

HAMILTON: *Departure Control, we've got an explosion. We have a flame. Oh, God. We've lost a wing.*

A few miles north of the airport flight 621 slammed into the ground and exploded killing 109 people.

Why did this happen? Was it the fault of the designer, the FAA, or a crew member? Although the final outcome is usually pilot error, and in this case the first officer did actuate the spoilers while still in the air, I still find it hard to explain why the FAA would certify an aircraft with such a glaring built-in booby trap.

One of the most effective means of learning is by experience, but the FAA does not even agree with this basic concept. If what you are about to read shocks you, wait until you read about the DC-10 dangers.

On August 7, 1970, a month after the Air Canada accident, the Douglas Aircraft Company, builders of the DC-8 and DC-10 aircraft, sent a cable cautioning all operators of the DC-8 to restrict the use of spoilers to ground operation only. No mention was made of the Air Canada accident. Before we continue, you should be made aware of the manner in which faulty situations are supposedly corrected by the issuance of various directives.

Services Bulletins: *These are issued by the aircraft manufacturer to notify carriers of recommended modifications. As the lowest priority bulletin, it is printed on white paper.*
Alert Bulletins: *These are similar to service bulletins except they denote a more important situation. It still is not a mandatory fix notice and it is printed on blue paper.*
Campaign Wire: *This is the third method by which the manufacturer alerts his customers. It can be sent by telephone, cable, or telegraph. These calls are then followed by an Alert Service Bulletin.*

None of the above directives have any mandatory force behind them. They may be heeded by the air carrier or ignored, with no reprisals from the FAA.
Airworthiness Directive (AD): *This is the muscle of the FAA, if they have any. The AD carries the force of law and compliance is mandatory within the prescribed time period.*
Telegraph AD: *This is the quickest method of dispatching a directive and it is used, or should be, whenever an emergency situation arises.*

Now that you are aware of the various degrees of enforcement of FAA orders, watch and see how reluctant the FAA is when it comes to issuing a lifesaving AD.

On November 12, 1970, Douglas advised the Western Region FAA that they had no record of a single instance when ground spoilers were ever extended in flight and that the proposed AD, emanating from the Western Region FAA, pertaining to placards in the cockpit and revisions to reflect this in company flight manuals would be completely unwarranted.

Not only did Douglas protest this proposed AD, but the ATA and Eastern Airlines followed suit. In spite of the protests on December 5, 1970, the AD was issued. This AD would be a relatively inexpensive item for the carrier, and probably why the FAA insisted on it. It consisted of additions to the aircraft manual stating: (1) "Do not extend ground spoilers during flight. (2) Install a placard on the flight deck immediately below the word 'Spoilers' to read 'Deployment in flight prohibited.'

That kind of sign is about as effective as one stating "Do Not Crash This Aircraft."

October 12, 1971, the International Air Transport Association, IATA, issued a safety exchange bulletin to its members pertaining to extension of ground spoilers in flight. Because of this second accident a directive was issued. It seems the captain of a DC-8 was having trouble activating the spoilers and by accident they were deployed fifteen feet in the air; this time only structural damage ensued.

November 28, 1972 a Japan Air Lines DC-8 crashed shortly after takeoff from Sheremetyevo International Airport, Moscow. Sixty-one of the seventy-six people aboard perished in the crash. The cockpit voice recorder was not too clear, but just before impact one of the Japanese crewmen said, "Sumimasen [Sorry]." The activation of all ground spoilers in flight remains a distinct possibility.

On June 23, 1973, 128 passengers were lucky enough to live through a landing they won't forget. Forty feet above the runway at Kennedy an Icelandic Airlines captain called for the spoilers to be armed. The F/O accidentally pulled the lever, and down came the DC-8. Thirty-eight were injured and the plane was damaged extensively.

It is interested to note just how much of an impression a cockpit sign can make. The Icelandic DC-8 was equipped with the AD issued by the FAA in December 1970: DEPLOYMENT IN FLIGHT PROHIBITED.

Following this latest mishap, the Eastern Region of the FAA drafted a proposed AD that would make activation of the spoiler in the air impossible. This worthwhile AD was sent to the Western Region of the FAA on August 6, 1973. Since Western Region, FAA, was in charge of aircraft manufacture in that territory, it was their duty to approve any AD in their area. Needless to say it was rejected.

Three years have passed since the Air Canada crash and the FAA is still playing games with safety. On September 13, 1973, Douglas informed Western Region, FAA, that it had engineered a foolproof lock-out mechanism so that in-flight activation of the spoilers would be impossible. They realized that a potential AD was in the offing so they decided to do something on their own. On October 5, 1973, Douglas issued Service Bulletin 27-251 to all DC-8 operators. This was a routine service bulletin, not even an alert bulletin. Douglas decided that this was a design improvement rather than—perish the thought—a safety modification.

Once again the Chief of Flight Operations, FAA, advised the Chief of Engineering, Manufacturing Division, FAA Headquarters, urging that a directive be sent out immediately to upgrade the service bulletin to an AD. He believed that four spoiler accidents with the DC-8 in the last four years should be incentive enough.

On January 15, 1974, four years after the first crash, the Western Region, FAA, finally got up enough nerve to issue an AD making service bulletin 27-251 mandatory. You may think that the FAA was finally doing something worthwhile, but there is a slight flaw in the AD. This modification must be performed before the next 3,600 hours of flight time, or about a year and a half hence. Probably enough time to have one more fatal accident before the fix.

The Jet Engine

The jet engine is far more efficient than the old reciprocating prop jobs, but it also has drawbacks that were not found in gas engines. If a gas engine blew a cyclinder, it would probably continue to function. Being of a more rugged nature it would most likely stay in one piece. The jet turbines turn so rapidly that a small foreign object, as small as a bird ingested into the intake could ruin a million-dollar engine.

Since the rotation of the turbine is so much faster (9,000 RPM) than a propeller-driven engine (2,000 RPM), there is a marked difference when a jet disintegrates. Pieces of blades and buckets are hurled against the fuselage with the force of shrapnel.

Since 1968 more than sixty failures of the Pratt and Whitney JT8D engines have occurred. These engines power the DC-9 Boeing 727 and Boeing 737. Although an AD was published and followed to the letter the failures continue.

June 13, 1972, an Eastern Airlines DC-8 had an engine disintegrate while on a training flight. The No. 2 engine sustained a 6-inch-wide circumference rupture of the turbine case. Engine fragments were shot through a tire, flaps and the wing.

On May 2, 1972, a Continental Airlines DC-10 was also on a training mission when the whole aft section of the center engine blew up. There was no injury inflicted by the engine, but all of these engine distintegrations are not to be so lucky. A massive failure such as this one has the potential to cause the loss of an aircraft by the damage the flying engine parts can do to the rest of the structure.

On November 3, 1973, Captain William R. Broocke was in command of National Airlines flight 27, a scheduled passenger flight between Miami, Florida, and San Francisco, California, with intermediate stops at New Orleans, Houston, and Las Vegas. The MACH meter was indicating a steady .82 at an altitude of 39,000 feet. All was serene until the plane passed over Albuquerque; then it happened. The passengers and crew were startled by a loud explosion and the aircraft began to buffet severely. Engineer Hanks saw the red glow of the light in the No. 3 fuel shut-off handle, and various other instruments on his panel indicated numerous system failures. Number 3 firewall shut-off was pulled and two fire extinguisher bottles were discharged into the failed engine.

Immediately after the initial explosion, large sections of the turbine blades were hurled through both remaining engines and foot-square holes were punched through the fuselage. The window at passenger seat 17H adjacent to the faulty engine was completely gone. The rush of air flowing outward from the cabin was equal to that of a hurricane. The passenger in seat 17H was sucked up into the open window in spite of his seat belt which he had loosely fastened. He was momentarily detained by his belt and by a fellow passenger who tried desperately to pull him inside the cabin, and for a moment it seemed he would be saved. But the tremendous pressure inside the cabin finally won and the passenger was completely sucked through a window that measured only 16 by 10 inches. The cabin immediately filled with smoke as is common with explosive decompression.

The passenger oxygen masks were supposed to fall automatically from their perches in the seat backs and ceiling, but only the masks in the mid-section of the aircraft were deployed. Other sections of the cabin had no oxygen masks for over three minutes. At 39,000 feet you have approximately one minute of consciousness. Two stewardesses in the lower galley passed out immediately. Many of the passengers began tearing at the closed oxygen compartments in an effort to remain conscious, but they received only severely burned fingers. A fire started in one of the seats.

Oxygen on the DC-10 is manufactured in the following way. Oxygen generators produce oxygen by thermal decomposition of sodium chlorate by a chemical reaction. These canisters can reach a temperature of $547^{\circ}F$.

The aircraft was nosed over in an emergency descent down to a level where oxygen would be plentiful. Radio contact was established with Albuquerque Approach Con-

trol and flight 27 was vectored for an approach to runway 26 at Albuquerque. The crippled plane was skillfully guided to a safe uneventful landing.

When the aircraft came to a stop, the cabin attendants opened the emergency exits and deployed the evacuation slides. The slide of the forward left door fell to the floor of the cabin and failed to inflate. The slide on the forward right door deployed normally, but also failed to inflate. All of the remaining slides operated normally except the one over the right wing which failed to deploy over the engine pylon, remaining instead on the wing where it was useless. Everyone, nevertheless, managed to get out through the emergency exits, and 127 grateful and frightened people were once again safely on the ground. Twenty-four passengers were treated for ear problems, smoke inhalation, and minor abrasions.

The one casualty was the unfortunate passenger who was killed when he was sucked out of the cabin window. I would like to interject a quote from an FAA official which will serve to highlight the caliber of the men at the upper levels of the FAA. This shameful quote is from an FAA propulsion engineer discussing the pros and cons of the engine that powers the DC-10:

"If the fragments strike and break a window, it is very likely that the passenger sitting adjacent to the window will be extruded through the opening; but here again, while we may have an unhappy *passenger, there is no diminution of the level of safety, and the airplane can continue to fly and land safely. One single missing window will not result in rapid decompression of the airplane."*

I wonder how this gentleman would feel if he were the passenger being "extruded." Also his statement that a broken window would not cause decompression is completely false. As we have seen, a broken window did cause this phenomenon, and what's more even a bullet hole would do the same thing.

On July 8, 1974, National Airlines Captain McGlannan

was climbing flight 41, a DC-10, through 25,000 feet. Suddenly a violent explosion accompanied by severe vibration rocked the craft. The plane was shaking so badly that the crew had difficulty reading the instruments. This time the No. 2 engine was the culprit. After three separate fire warnings were observed, it was shut down and the fire bottles were directed to it. An emergency descent was started and an uneventful landing was accomplished on runway 36 at Tampa, Florida. None of the 172 occupants was injured and all passengers deplaned at the gate through the normal exits.

Inspection of the aircraft revealed that the left engine cowling was missing and there was a 9-by-9-inch hole in the upper left wing. The middle engine was damaged extensively, both internally and externally. This time it was not the fault of the engine; sloppy maintenance was the culprit.

A previous pilot had complained that the left engine reverse lock light remained on after landing at Miami. Maintenance had removed the No. 1 engine cowl in order to replace the engine reverser. After the component change, the cowl was replaced with only one bolt holding it. The mechanic who was working on this cowling gave a verbal notification to the man on the next shift, and he went home. This oncoming mechanic was then moved to another aircraft, and when he returned, he assumed that the cowl was properly fastened.

This unsafe cowling had been supervised by line maintenance during five work shifts. The aircraft was needed for a scheduled flight, so, on July 8, 1974, 160 passengers and a crew of twelve got a thrill they could have done without.

In December 1973 a KLM DC-10 was on its takeoff roll when the center engine let go and threw blades at various vital components of the engine. One blade cut a fuel line and a fire ensued. The pilot was able to abort the takeoff and extinguish the fire thereby averting a potentially serious accident.

The General Electric CF-6 jet engine that is used on the DC-10 and the European sky bus has a unique feature not found in other jets—each blade of the turbine wheel has sixteen holes drilled in it to decrease weight and increase acceleration potential.

Evidence has been accumulating substantiating the fact that these drilled blades are highly susceptible to foreign object damage (FOD) from gravel, birds, ice, etc. There are over fifty-eight cases on record where blades have been damaged, and in many instances, have been flung through the engine casing rupturing the aircraft. Solid blades are not nearly the hazard that hollow blades are, but unfortunately, the majority of the DC-10 aircraft have the drilled blades.

In 1970 the FAA evaluated the CF-6 engine. It ran a flight into icing conditions to check out the blades. The CF-6 engine is the first of the big engines that have no anti-icing system. On September 16, 1970, the Western Region FAA, issued a type certificate to the CF-6 engine.

There were questions raised by the FAA Headquarters and others that the icing tests were not adequate because the craft had only encountered light icing. So, on March 28, 1971, more icing tests were conducted (after it was already certified by Western Region). These tests proved conclusively that the drilled blades were not as rugged as the solid blades. After a test flight of one hour, 16 minutes in light-to-moderate icing conditions, three blades of No. 1 and 26 blades of No. 2 engines were severely damaged.

On March 29, 1973, the A300B airbus was undergoing icing tests in France and both engines were damaged. The FAA representative in France sent the following letter to FAA, U.S.A.: "To preclude damaging the fanblades when operating in icing conditions, we will require as a condition of A300B certification, that its CF-6-50A engines have only the undrilled blades."

A wire from FAA in the States was quickly dispatched to Brussels demanding a retraction concerning the engine blades, because it would have an effect on the already certified engines in the States. When the boss calls you obey or else! So, on February 13, 1974, FAA Brussels rescinded their letter requesting only solid blades.

There were numerous letters and meetings between the FAA, NTSB, Douglas, and other interested parties all raising various questions concerning the three potential bombs attached to the DC-10 aircraft. On August 23, 1974, the FAA issued an AD requiring all DC-10 aircraft to have a protective shield around the center engine's fuel lines. This was to keep the turbine blades from puncturing the fuel line and hopefully to prevent another accident.

It is interesting to note that the FAA would rather protect the vital parts of the engine from blade shrapnel than change to solid blades.

Well, this is another in a long line of ADs that did not stick. Due to obvious pressure, the FAA rescinded the AD stating that there was really no unsafe condition at all.

In October 1974, a subcommittee chaired by Representative Harley O. Staggers of West Virginia questioned the operations of the FAA and dug deeply into the problems of the DC-10. He wrote a letter to the Hon. John H. Reed, Chairman of National Transportation Safety Board, in which he stated the following: "April 18, 1974. A key official in the Western Region, FAA, stated that 'Service and test experience indicate that the drilled GE CF-6 engine fan blade is very sensitive to FOD. The resulting damage to the engine and aircraft is considered excessive and presents a serious threat to the safety of the aircraft.' "

February 15, 1974. FAA Chief of Aircraft Certification Staff in Brussels.

"Advise FAA Headquarters that A300B airbus is particularly vunerable to FOD because both engines are subjected to the same operating environment and icing conditions." He noted that damage to a second engine "Could be disastrous."

May 14, 1974. The chief propulsion Branch FAA Headquarters "Aircraft damage from shrapnel trajectories forces can be such as to puncture cowling, damage fuselage and disable systems."

During Rep. Staggers subcommittee investigation which lasted for days, there was a wealth of good testimony given, but it took a great deal of patience on the part of Rep. Pickle and others. What follows are some of the questions posed by Rep. Pickle and answered by the FAA.

MR. PICKLE: *The FAA has not issued an AD on this particular engine and you do not classify the engine particularly with reference to the drilled blades as being unsafe; you do not consider this an emergency situation. Is that correct?*

MR. BUTTERFIELD: *That is correct; we do not consider it an emergency situation.*

MR. PICKLE: *Because you do not consider it an emergency situation, you have not issued an AD?*

MR. BUTTERFIELD: *The emergency type AD would ground the fleet.*

(At this time there was a proposed FAA rule to make all carriers change from drilled blades to solid ones.)

MR. PICKLE: *This is very important now for us to understand you. Are you saying to us that you, the FAA, might withdraw that proposed AD?*

MR. BUTTERFIELD: *No, I am not saying that.*

MR. PICKLE: *Why did you mention it, then?*

MR. BUTTERFIELD: *I am saying it is not a directive.*

I wanted to make the point that we are not directing the industry to do anything at this time.

MR. PICKLE: *But it is your interest to issue such, is it not?*

MR. BUTTERFIELD: *Yes, it is.*

A few months after the above hearing Mr. Butterfield made a complete reversal regarding the proposed AD. The following is a quote from his letter to Rep. Staggers: "Therefore the FAA has withdrawn the Notice of Proposed Rule Making that would require installation of solid fan blades. Solid fan blades will continue to be installed on an optional basis."

On June 12, 1972, American Airlines flight 96, a DC-10, was climbing out of Detroit's Metro Airport. First Officer Pete Whitney, age 34, was at the controls, under the watchful eyes of 52-year-old Captain Bryce McCormick. The cabin was being supervised by Chief Flight Attendant Cyda Smith aided by seven assistants. The fifty-six passengers were relaxing and just settling down with a good book, a martini, or both.

Flight 96 was cleared to climb to 21,000 feet. The altimeters were showing a rapid rate of climb and as the plane passed 11,750 feet the crew was startled by a thud which they both heard and felt.

CAPTAIN: *What in hell was it?*

FLIGHT ENGINEER BURKE: *We've hit something.*

WHITNEY: *We've lost an engine here.*

CAPTAIN: *Which one is it?*

BURKE: *Two.*

CAPTAIN: *Is pitch hold grabbing on?*

BURKE: *Ah, ah, Captain. . . we'll have to check this out.*

WHITNEY: *Do we have hydraulics?*

CAPTAIN: *No. I've got full rudder.*

WHITNEY: *Ok, we've got full control on this though, however, I guess we're slow enough so we can probably use differential directions with engines. Thank goodness it's one and two, even got one and three.*

CAPTAIN: *OK, give me about fifteen on the flaps now, watch it carefully.*

BURKE: *We'll be landing about 292,000 pounds.*

(I've taken the liberty of excluding radio conversation for clarity. Only the conversation in the cockpit is heard.)

WHITNEY: *We've got a nice rate of descent. Even if we have to touch down in this way we're doing well.*

CAPTAIN: *Well, give me the gear. They're OK here. We're coming into the ILS. I'm going to start slowing her down. Give me 22 on the flaps.*

(Now the crucial part of the whole flight would be unfolded in the next few seconds, the touchdown and rollout.)

CAPTAIN: *I have no rudder to straighten it out with when it hits.*

WHITNEY: *OK. Engines off at your discretion.*

CAPTAIN: *Shut 'em down.*

The crew performed admirably. To bring one of these giants down safely with no rudder control and elevators so stiff it took the strength of both pilots to raise the nose on touchdown is quite a job. All this with no injuries whatsoever.

The cause of this near disaster wasn't a bad engine, a bomb, or loss of hydraulic pressure; it was caused by what appeared to be an insignificant item—an improperly closed rear baggage door. It may sound like an unimportant matter, but because of the design of this aircraft, the baggage door is of paramount importance.

The ramp service agent who loaded the aft cargo compartment had difficulty closing the door. He stated that he was unable to close the handle with normal force, so he put his knee on it to force it closed. The door seemed to be shut, but it was not latched properly inside. Although the inside lock pins were not in place, there were no warning lights on the engineer's panel. When the aircraft was climbing for altitude, the high pressure in the cabin blew out the cargo door. Then the floor of the aircraft over the compartment buckled and a number of important flight and engine control cables jammed. The DC-10 is the only wide-bodied jet that has these most important cables traversing the floor. All others have theirs in the ceiling.

This time a potential disaster was averted through a combination of skill and luck. This accident sets the scene for the worst disaster in aviation history.

On July 6, 1972, the NSTB issued a recommendation to the FAA to modify the cargo door locking mechanism. Acting within its own authority, the Western Region of the FAA prepared to issue a telegraphic AD requesting a functional check of the door by all carriers. A bulletin prepared by Douglas would also be incorporated in the AD.

Service bulletin 52-35 which was to be included in the above AD was issued by Douglas by telegram on June 16, 1972. It called for the installation of a one-inch-diameter window on all three lower cargo doors. This would enable the baggage handler to look behind the door to ascertain if the lock pins were properly closed (that is if he were to look at all). It also called for three instructional placards to be affixed to the door.

The Western Region three-man review board approved the AD on June 16, 1972, four days after the Detroit incident. All that remained was the signature of the Western Region Director Mr. Basnight. At last it seemed as though this important AD was really going to become a reality . . . but not quite so fast.

On the evening of June 15, 1972, the FAA Administrator John Schaffer telephoned Mr. McGowen, President of Douglas Aircraft, and asked what they at Douglas were doing concerning the cargo door. Mr. McGowen said that they had a proper fix and there would be no problem. Mr. Schaffer replied that the corrective measures could be taken without an AD. That sounds more like the FAA I know. In lieu of this important AD, Washington FAA elected to call the presidents of airlines using the DC-10 and request that they have their cargo doors modified. This still is not mandatory, only an AD is.

Douglas was preparing a fix and part of it consisted of increasing the pressure required to close the door handle from 110 pounds to 440. This they said would make it virtually impossible for the door to be closed improperly.

A service bulletin was issued July 3, 1972, involving voluntary modification of the cargo door. This will give you an idea of how rapidly the airlines comply with these service bulletins: of the thirty-nine DC-10s in domestic service, five were modified within ninety days, eighteen were not modified until June 1973, and the remainder not until much later. That's over a year after the SB was initiated.

Aircraft manufacturers naturally take pride in their product and well they should. It takes years of planning and testing before the first production model runs off the assembly line. But they should be taught that a spade is a spade, especially where human life is concerned.

The aircraft manufacturers believe their products are just about perfect, and they will promise to correct any airworthy item, if only the FAA will refrain from slapping an AD on them.

During a recent Congressional subcommittee investigation concerning the DC-10, there were a number of pertinent statements made.

Representative Pickle's questions to Mr. Pittman, Engineer WR, FAA:

MR. PICKLE: *I would like you to state again, now, how you view an AD, and how you think management views an AD? I didn't get a close picture of your feelings and opinions. Will you restate that?*

MR. PITTMAN: *Let me state how I feel. The manufacturer looks at it, at this point, or what I understand from what they have told me, an AD, for one thing, has been mentioned as a sort of black mark on their product, or at least they feel so.*

Representative Lent's questions to Mr. Pittman:

MR. LENT: *Mr. Pittman testified that he had prepared numerous ADs in his years with the FAA. Can you tell us of the ADs that you personally prepared and worked on? How many of these were killed after they passed the AD Review Board stage?*

MR. PITTMAN: *None, sir.*

MR. LENT: *So this is a rather unusual circumstance, as far as you are concerned?*

MR. PITTMAN: *For me, personally, yes.*

MR. LENT: *In all your years with the FAA, this is the first time that some outside branch of the FAA has countermanded a regional AD?*

MR. SLIFF (Chief Engineer, WR, FAA): *To my knowledge, that is true.*

Now Representative Manellie quesstions Mr. Sherman, FAA, project engineer for the DC-10.

MR. SHERMAN: *It is my opinion that the people that we generally deal with, at the engineering level, probably favored the AD. However, I would be under the impression that perhaps management would be less in favor of an AD.*

MR. MANELLIE: *Would one of the reasons why this might be true be that the competi-*

tion would sometimes use ADs as an argument that their product might be better than the competitor's product, which has an AD issued on it?

MR. SHERMAN: *That might be possible.*

On March 3, 1974, the worst tragedy in the history of aviation took place in Paris, France. Shortly after takeoff, a Turkish Airlines DC-10 plummeted to earth, killing 346 people. This craft was at almost the identical altitude as that of American's in the Detroit incident. While the exact cause has not been determined, the experts are of the opinion that the plane suffered an explosive decompression. A door of the plane was found more than six miles from the main wreckage . . . a rear cargo door.

Three days after the crash, Washington FAA, finally issued the much talked about AD. It is of interest to note that the very officer who squelched the issuance of this AD by Western Region, FAA, was the one responsible for the issuance of this directive.

The Turkish DC-10 was in compliance with Service Bulletin 52-35, which included the inspection hole and an instrumental placard. Unfortunately, the placard containing the proper method of closing the cargo door was written in English, and the cargo handler in Paris was French. Simply because an inspection hole is put on a door is no guarantee that the person closing it will look through it.

On March 14, 1974, eleven days after the accident, the FAA ran a survey to see how many carriers were experiencing trouble shutting the cargo doors. They discovered there were over 167 discrepancies found in aircraft logs. Running the gamut from a door that was simply hard to close all the way up to pilots who had to return and land because of the door seal was leaking so badly they could not pressurize the aircraft.

The FAA claims that this crash can in no way be connected to them. They put the monkey entirely on the back of Douglas. It seems that the Turkish DC-10, along with another that was delivered to Laker Airways, was still being built at the time of the Detroit incident. The latest door mods were supposedly installed at the factory. In fact, the Douglas Aircraft inspector initialed that the work on these two planes was completed, which was not the case.

That the AD should have been issued after the American incident is now a moot point. If this AD was implemented, as many of the FAA engineers requested, it would most likely have ended up at the engineering offices of the Turkish airline. This would have given them a chance to check the door and find that its mechanism was defective.

The following is a list of airworthy items that were all going to be remedied by an AD:

Lockheed 1011 rudder pedals were jamming, causing serious control problems.

Lockheed 1011 auxiliary power unit exhaust shroud seal was a potential fire hazard.

Lockheed 1011 passenger oxygen masks failed to deploy when needed.

Boeing 747 escape slides failed to operate normally, endangering the lives of passengers.

All of the above, and many more airworthy items, were being considered for ADs, and then the manufacturers persuaded the FAA to cancel them, as they would take care of the matter themselves.

Well, I guess we've all seen enough of that.

Most people believe that the FAA has only two main functions—to make and enforce regulations. However, they also have the responsibility to foster and promote aviation. Herein exists a conflict that should be changed, as it is nearly impossible to do both.

A recent subcommittee representative, Mr. Carney, was questioning the head of the FAA, Mr. Butterfield:

MR. CARNEY: *Where do you think your prime responsibility lies?*

MR. BUTTERFIELD: *Safety, without question. And there has been greatly increased emphasis on the subject over the past year, as I believe you have noticed.*

MR. CARNEY: *I'm glad to hear you say that, because my impression from the two cases I sat in on is that the safety of the public comes second to the welfare of the industry.*

I am in full agreement with Mr. Carney. By now, I hope you are, too. However, I must take issue with Mr. Butterfield's statement concerning safety last year. 1974 was the worst year, safetywise, in the last ten. How many more lives must be lost before the FAA comes to its senses.

Chapter Twenty

WHY DOESN'T THE FAA MAKE IT MANDATORY?

WITHIN the next few years, airline passenger miles flown will increase from approximately 190 billion miles (1968) to 500 billion miles (estimate for 1980). It is estimated that during these years some 4,000 passengers will be killed by aircraft fires unless there are drastic changes made to evacuation and fire-suppression systems.

A battle has long raged among the world's airlines regarding the safest fuel. Some users favor straight kerosene, known as Turbine A fuel, and others prefer a mixture of gasoline and kerosene called JP4. Laboratory tests have shown that straight kerosene is the least volatile. Nearly all U.S. carriers use kerosene and only a few foreign airlines favor JP4 which is slightly cheaper and weighs 1.4 pounds less per gallon than kerosene. Regardless of the fuel used on an airplane, it is still vulnerable to lightning strikes and the danger of explosion during ground emergency.

In December 1963 a Pan American 707 holding over Elkton, Maryland, blew up after being struck by lightning. Eighty-one people died. In November 1964 a TWA 707 crashed after an aborted takeoff. Minutes later, when it came to rest, an explosion occurred and killed over double the amount of people originally lost in the crash. There are many more cases on record where the explosion after landing was the real culprit.

Let's explore the three ingredients necessary to produce combustion: fuel, oxygen, and an ignition source. If any of these ingredients are missing, fire will not take place.

There is available today a tried and proved depressant that will eliminate gas tank explosions: fuel inerting. This system replaces the oxygen in the fuel tanks with liquid nitrogen. To achieve this protection it is not necessary to eradicate all of the oxygen, only enough to arrest the propagation of the flame. If the oxygen content is kept below 10 percent there will be no explosion.

As the fuel in the wing tanks is sloshed around during flight, the dissolved gases come out of solution in an uncontrolled rush. The fuel tank can be likened to a giant soda-pop bottle. Unless the bottle is agitated the gases stay in solution. But as soon as you shake the bottle, out come the gases.

The only action required to make the inerting system operational is to service the aircraft with nitrogen. Once the liquid nitrogen is on board the plane, it is fully protected, even with the aircraft unattended and all electrical power turned off. Should a fire in flight become a reality, the pilot has merely to press a button and the inerting gas is sent to the affected area. Nitrogen is nontoxic and noncorrosive. The cost of this extinguisher is three cents per pound; a typical flight may consume 300 pounds of LN2 and would thus cost approximately $10.

The U.S. Air Force tested this system on two of its transports, the C-135 (similar to the 707) and a huge C-141 cargoliner. During the tests over 100 hours of flawless operation was logged with no maintenance required. The Military Airlift Command was so enthused with the system that an order was placed to have their entire fleet of C-141s nitrogen inerted. The added weight penalty required for this system is approximately

700 pounds for a 707 and 1,300 pounds for a 747. These figures include the supply of nitrogen. The initial cost, if installed while the craft is being built, is less than the cost of in-flight entertainment.

"We must face the fact that the smallest airplanes flying can cause the destruction of our biggest and finest airliners and all they contain as a result of simple collision. As size and number of airplanes increase, the problem becomes more acute." That statement was made in 1953, at ALPA's first Air Safety Forum. It is disturbing to note that, nineteen years later and with very few changes made, we are still pressing for a safer system. The "see and be seen" concept has been with us since the government decided to make some rules for the boys upstairs; it seems only fitting to give the pilots a fighting chance to see each other.

The electronic strobe (similar to a camera flash gun) has been around for years. It is a necessary safety item, especially in the hazy crowded skies near large airports where seeing the other fellow in time is the only chance for survival. With these strobes mounted on each wing tip, traffic can be observed at a great distance in the daytime, and in the evening the range is phenomenal. These lights not only save lives, they save money as well.

The bird-strike records of two comparable airlines flying similar equipment to identical airports was closely examined. It was found that in the year 1968, the airline with the strobe lights experienced only one bird strike and repairs cost $1,695. The carrier without the lights had to pay over $175,000 to repair the damage caused by our feathered friends, not to mention the revenue lost by having a plane out of service. It is an obvious assumption that birds stay clear of these flashing monsters.

At least in the old days a DC-3 pilot on a collision course could flash his landing lights in the hope that they will be seen by the other crew. The lights were permanently facing forward and could be used for this purpose at any speed. Lights on a jet retract into the wings and cannot be lowered until well below cruising speed. It does seem disgraceful that the jet age captains must still signal one another to avoid collision, just as the ancient mariners did as they passed in the night.

Once again the military are better equipped than the civilian fleet when it comes to a very important piece of equipment: the automatic pilot. Most of the government's fleet of jets have this device, and they seldom carry passengers. Yet there is no regulation that requires an automatic pilot on board a commercial airliner. Some airline executives believe that an autopilot is a luxury item. One senior VP made this statement: "I pay pilots to fly my planes. Let them earn their money." Well, some people are not aware of the tremendous workload of a pilot. "George" gives a man a rest when he needs one. All airlines should have automatic pilots, especially the puddle-jumping carriers with only two men up front.

The FAA and the company manuals clearly spell out the procedure to be used when turbulent weather is encountered:

1. Autopilot *on*.
2. Altitude hold *off*.

If there is no autopilot on board, how can this be accomplished?

The radar controller, like the pilot, is greatly overworked. Any aid that will make his job easier will in turn add safety to the system. A transponder can make life worthwhile on a clear day. It becomes a necessity during foul weather as that is when the traffic piles up. When a controller is not sure which target he is talking to, he simply requests the pilot to "Squawk ident." All a pilot has to do is to push a button. The target is now readily seen on the scope because its size nearly doubles.

This transponder is required only above 18,000 feet and in the vicinity of certain airports. Since the majority of the planes in the air are private and cruise well below that altitude they are not required to have the "black box." Yet transponders are available for light planes for under $300.

The "state of the art" concerning pressure altimeters has

reached the high point and they are still unreliable. The radio altimeter which sends a beam down to the ground and back to be registered in the cockpit is far more accurate than the barometric that is used mainly today. There are infinitely less variables to cope with when using this superior altimeter, and it would be a good backup for the ancient ones in standard use now. The only regulation making it mandatory for the radio altimeter to be on board is for the limited number of aircraft supposedly equipped to allow lower landing limits.

If you were to visit the cockpit of a 707 that belonged to American Airlines, you would find a completely different arrangement than on a 707 belonging to TWA. The airlines are at liberty to have their cockpits arranged as they choose, but what happens when one line sells another line one of its planes? Chaos! Most of the time little, if anything, is done to standardize the new plane's cockpit to their present fleet.

Picture, if you will, an airline that buys its planes from many different carriers. Now the pilot is required to memorize and be fully cognizant of several completely different cockpit layouts. Here are some of common cockpit discrepancies: switches in different places, control handles not where they should be, and knobs to control the engines within four inches of similar ones that will instantly stop them. Add to that instruments that cannot be seen unless contortions are practiced.

British aircraft such as Viscounts and BACs have switches that operate in precisely opposite directions to American-built planes, radio knobs that have no definite feel as to whether they are on or off, vital switches that are not guarded, circuit breakers that follow no set pattern, power settings that must be checked for each model, instruments that have different scales for the same gauge. One plane may have an altimeter with three pointers; the next plane you fly has an altimeter with one. Even the paperwork concerning the allowable weight for landing may be different if similar planes have unlike engines.

The U.S. Air Force is cognizant of the fact that it is a full-time job to master one type of aircraft, and that is all their pilots are allowed to fly. But commercial airline pilots are still permitted to fly two or more totally different types of aircraft on the same day. A highly confusing, frustrating, and unsafe experience. Still, there is no law against it.

Chapter Twenty-one

THEY HAVE THE POWER

"PROBABLE CAUSE" is a phrase that has been used extensively throughout this book. How does the Supreme Court of Aviation, the National Transportation Safety Board (NTSB) arrive at the cause of a crash—a decision that must be shared with the wife and children of a pilot who allegedly made a fatal mistake?

Prior to 1966 the Civil Aeronautics Board was responsible not only for aircraft accident investigation, but also for the regulation of airline fares and the awarding of routes. It was a clear case of too much work for too few people. In 1966 the NTSB was formed. At last the accident-investigation field would have the undivided attention of a panel of experts. Or would it?

NTSB Chairman John H. Reed, former governor of Maine, had this to say regarding the new board: "For the first time there is a federal agency completely dedicated to the safe transportation of people and products. It has only one mission, to improve the level of transportation safety."

The National Transportation Safety Board was created by

Congress in 1966 when it passed the Transportation Act of that year. While the board is affiliated with the Department of Transportation, it functions as an autonomous agency. The act states that the board "in exercise of its functions, powers, and duties shall be independent of the Secretary [of Transportation] and other officers of the department." The board consists of five members appointed by the President with the consent of the Senate. In the hierarchy of federal departments in Washington, the NTSB is one of the smallest agencies, with about 250 employees.

The Bureau of Aviation Safety, which was inherited from the CAB, is directed by Charles O. Miller and supplies the technical knowledge so necessary in determining the cause of a crash. This bureau is similar to the FBI in one respect: its engineers trace the tiniest clues that may have any bearing on a crash.

The main home of the NTSB is Washington, D.C., with eleven small field offices around the country. In carrying out its mission under the Transportation Act the board determines the cause of all airline and fatal light-plane accidents no matter where they occur. Accidents to foreign carriers are also investigated if they fall on U.S. soil. At headquarters in Washington the NTSB maintains a Go-team that is in constant readiness to speed to the scene of disaster. It consists of ten aeronautical specialists who work at regular desk or laboratory duties daily and are on standby to fly to accident scenes. One member of the five-man board is assigned to this duty also, and he remains on "ready" status for periods of seven days.

The board is unique in several ways. First, the safety recommendations the board makes are not mandatory. The board is neither an enforcement or regulatory agency—and that's the way they want it. Second, the board must make public all of its recommendations. This is their big stick to shame the offenders into proper safety procedures. Enforcing new regulations is the job of the lip-service FAA.

One of the greatest assets in determining the cause of an accident is the aircraft's flight recorder. This unit is required in all jet airlines. Encased in a tough outer shell, the flight path of each airliner is automatically traced on a thin strip of aluminum foil. Such vital data as airspeed, altitude, direction, "G" forces, and time are all clearly recorded. Another telltale clue is the cockpit voice recorder. This automatic tape recorder is turned on before each flight taxis out. It records all the voices of the crew in the cockpit as they are performing their duties and it also picks up the radio calls and cabin announcements. These two units are a boon to air-crash detectives. Unfortunately, in many catastrophes they are destroyed by fire and along with them the guilt or innocence of the crew.

The board's accident reports would seem to the average person to be clear, concise, and thorough. Once a formal report is issued it is bound in a blue cover. Reports average about thirty pages in length. There are numerous graphs and cross-sections of the fateful flight and a complete history concerning the experiences of the crew and the log of the aircraft. There are, however, an overwhelming number of pilots, most of them dead, blamed for the various crashes. This is one way to get the case off the books; a dead pilot makes an excellent scapegoat.

On November 19, 1969, Captain Raymond P. Hourihan, thirty-one, was in command of Mohawk Airlines Flight 411. A twin-engine turboprop Fairchild operating from LaGuardia to Glens Falls, New York, with a stop at Albany. The flight was relatively uneventful until the last leg to Glens Falls. Captain Hourihan exceeded his clearance limit and for some reason failed to descend for his nonprecision VOR approach. Instead he continued 9 miles past the airport and slammed into a mountain, killing all on board.

The board's probable cause in essence placed the entire blame on the captain. Let's take a thorough look at some of the board's statements. First, aircraft information concerning the logbook write-ups: "1.6: Further investigation, however, revealed that these problems did not constitute *a trend* in navigation instrument problems."

The aircraft logbook showed the following flight instrument and navigation squawks:

9/10/69. Co-pilot's VOR navigation shows station passage when in reality nowhere near it.
Write-off: Replaced co-pilot's VOR receiver.

9/11/69. Co-pilot's VOR still showing erroneous station passage.
Write-off: Reseated the VOR receiver.

9/12/69. Co-pilot's VOR indicator intermittent on some frequencies. Momentary erroneous station passage.
Write-off: Replace VOR receiver and checked with SG-13 tester.

9/21/69. Captain's course indicator sticks.
Write-off: Replaced indicator.

10/9/69. There is a 4 degree discrepancy between captain's and co-pilot's compasses.
Write-off: Cleaned and reseated course directors.

10/12/69. No. 2 compass reads 20 degrees high.
Write-off: Compass has returned to normal.

10/18/69. Instrument used to control pitch and roll erratic.
Write-off: Replaced vertical gyro.

10/24/69. Captain's VOR shows 8 degrees difference from co-pilot's.
Write-off: Reseated VOR.

11/3/69. Co-pilot's VOR erratic needle bounces.
Write-off: Replaced VOR receiver.

11/9/69. No. 2 gyro slow to erect.
Write-off: Replaced gyro.

On November 14, five days prior to the crash, Captain John Crumblish was in the holding pattern at Binghamton, New York. The weather was bad, and as he made his first turn outbound away from the fix his compass card began to rotate rapidly. He leveled the wings and the compass straightened out. Each time he turned outbound, the compass card rotated rapidly, making it useless. He was fortunate enough to be working a tower that had radar. He asked for a vector to a nearby airport where weather conditions were such that he could make a visual landing. But Captain Hourihan was not lucky enough to be landing at an airport with radar, or for that matter even a tower.

11/14/69. Co-pilot's VOR erratic. Captain's compass shows wrong heading in turns.
Write-off: Changed inverter. [This converts DC to AC power which is the force that drives the flight instruments.]
11/15/69. Co-pilot's VOR failed.
Write-off: Replaced VOR.
11/17/69. When operating on the spare inverter both VOR failed.
Write-off: Replaced the spare inverter.
11/18/69. DME [distance measuring equipment] inoperative.
Write-off: Checked OK on test.

This aircraft obviously had a history of flight instrument and navigation system problems. For seventy days prior to the accident the aircraft averaged a vital instrument discrepancy every four days. As the flight was operating in bad weather and at night, these instruments were vital and yet the NTSB says that these problems *"did not constitute a trend."*

The board stated on page 16: "The inverter was finally replaced and this proved to have been the problem." This is in error; the discrepancies still existed *after* the inverter was changed.

Power plants are usually returned to the manufacturer to be checked for malfunctions. ALPA repeatedly requested that this be done. But it never was. The Rolls Royce Dart engines that powered this craft were so designed that if a negative "g" force was experienced for a short duration one of the engines could automatically shut down. This plus the fact that two independent witnesses testified to the possibilities of an emergency situation on Flight 411 should have prompted engine teardown.

Months after the accident, the engines were removed from the crash site to a local junkyard—unexamined.

The board stated: "Three witnesses in the Glens Falls area saw the aircraft as it proceeded northbound. Therefore it is believed that the crew of MO-411 at least could have seen the lights of the town."

With a malfunctioning navigation system and flying in turbulence and rain, a crew would not be looking out the cockpit windows. The wind reported at 411's flight altitude of 3,000 feet was averaging 61 knots. And the reported cloud height was 2,000 feet overcast. It would be difficult for any crew to be able to see through 1,000 feet of solid clouds to the ground.

There was no public hearing concerning the accident, contrary to the usual procedure. In the NTSB annual report to Congress in 1969 there appears an interesting statement concerning the case of a light plane crash: "The Board conducted a major accident team investigation and held a public hearing *to develop the fullest possible factual record on the case.*" Why was there no public hearing in the case of Captain Hourihan? If a famous statesman or movie star had been on board Captain Hourihan's plane, the press coverage would have triggered an extensive, precise investigation.

The president of ALPA sent a critique to the chairman of the NTSB, John Reed. In it he questioned many of the NTSB's statements and asked for a joint meeting between the ALPA representatives who took part in the investigation and the safety board. Safety chairman Reed's answer was completely negative, and on page 3 of his letter there is a rather startling statement: "In conclusion, *it is not essential whether we are precisely correct in the analysis of the facts, or the ultimate determination of cause* but that we have stated all the facts fairly, fully and accurately and that the ultimate conclusions are reasonable and supported by these facts" (italics added).

It is quite evident that the board lacks the remotest trace of inquisitiveness. The pilot of Flight 411 exceeded his clearance limit and was also 9 miles from the airport in the wrong direction. But why?

In December 1968 an Allegheny Airlines turboprop Convair crashed while making an instrument approach at Bradford, Pennsylvania. On January 6, 1969, a similar accident happened to the same airline at the same airport.

In deciding the probable cause of the first accident the blame was placed on the deceased pilots: they were looking out of the cockpit window for a glimpse of the runway and simply flew into the ground. Captain Hourihan of Mohawk was criticized for not looking out to see the airport. You're damned if you do and damned if you don't. Regarding the second Allegheny crash at Bradford, the board made a concluding statement that is far more apropos than many of their other probable causes: "The Safety Board is unable to determine precisely the probable cause of this accident." It takes far more courage to make that statement than it does to blame the innocent.

In none of these crashes did the board find any discrepancies regarding the then minimal airport and approach facilities. Both Bradford and Glens Falls have minimum-length runways, no approach lights, no VASL, no control tower, no radar, and no precision facilities. There was no electronic beam (ILS) to guide the pilot for a safe descent. All these accidents happened at night in adverse weather conditions.

It's about time the monkey was put on the back of the people who foster the numerous approach accidents: the FAA. By allowing scheduled carriers to operate into substandard death-trap airports, they invite disasters.

Captain Vernon Lowell of TWA has long been a crusader for safety. His book *Airline Safety Is a Myth* is living proof of his dedication. As one of the ALPA official crash investigating officers assigned to the following accident, Captain Lowell had an inside view of the NTSB in action.

On November 17, 1967, a TWA Convair 880 crashed while making a nonprecision instrument approach into rain-swept Cincinnati airport. Seventy persons lost their lives. The board stated that the pilot was busy looking for the runway and lost sight of his altimeter. Where have we heard that one before? They said: "The approach was conducted using visual reference to partially lighted irregular terrain which may have been conducive to producing an illusionary sense of adequate terrain clearance."

Captain Lowell was so riled by this report that he sent a letter to Transportation Secretary John Volpe. He explained the inadequacies of the report with accurate and thought-provoking statements.

Prior to their accident TWA had modified the static ports, where the altimeter gets its input, on the fleet of 880 aircraft. This was done by enlarging the center hole so that it would accommodate their testing equipment. This modification was approved by the FAA, Kollsman Instrument Co., and the Convair Aircraft Company. These enlarged holes made water ingestion more prevalent with the accompanying potential of freezing, rendering the altimeter, airspeed indicators, and vertical speed gauge inaccurate.

Captain Lowell wrote: "The Board disbelieved and discounted a report by the airline [TWA] which stated that the information the pilots were seeing on their instrument panel was not what the aircraft was actually doing. Why should the Board discard this expertise?"

NTSB's annual report to Congress in 1969 had this to say concerning altimeters: "The reassessment of altimetry systems with particular regard to their susceptibility to insidious interference by forms of precipitation needs to be the subject of attention by the highest level of aeronautical facilities and personnel."

After the accident TWA went to the expense of installing static port heaters, which points up the fact that the airline believed the trouble was caused by these ports.

The board further stated: "The cockpit crew conversation reflects a relaxed atmosphere in the cockpit until the last few seconds before impact."

I am sure that neither you or I would relish the thought of flying in a plane in which the cockpit has a tense atmosphere.

Captain Lowell's letter continues:

There was another very interesting aspect of this accident. Several days before the public hearing in Cincinnati another Convair 880 descended too low and flew through some electric wires at O'Hare Airport in Chicago.

The meteorological conditions were similar to those in Cincinnati. The glide slope was inoperative. At the pre-hearing conference in Washington, D.C., the Board established some ground rules for the Cincinnati crash hearings. One rule was that we could not discuss any other accident. This was most unusual, since another similar accident had just happened in the same type aircraft with modified static ports, under similar weather conditions (precipitation and temperature) and on an ILS approach with an inoperative glide slope.

At the public hearing in Cincinnati the Chairman promptly struck from the record any mention of the O'Hare accident and warned me not to mention it during the interrogation of witnesses. Why would he do this? That is, if the Board really wanted to determine a true probable cause.

According to the rules established by the board on the procedures to be used at accident hearings, the chairman in the above mentioned situation was not following the book. Part 431 Rules of Practice in Aircraft Accident Investigation Hearings Section 431.2 has this to say: "During the course of the hearing no objections to any matter will be entertained from any part to the investigation or any other person."

Captain Lowell's letter to the secretary contains a paragraph that is so well put that I must quote it in toto.

> Statistics aside, many lives would be spared if accurate determination of the true probable causes of airline accidents were made and correctly reported. I am not criticizing the technical groups involved in the investigations as I feel they perform quite well. [That is debatable.] However, I must denounce in the strongest terms possible the NTSB's accident reporting. Too often their reports are distorted, inaccurate, amateurish in their technical aspects and, yes, even dishonest.

This fine letter, after being sent to Mr. Volpe, was re-sent by the secretary to the board for them to answer. Said Francis H. McAdams, one of the most senior members of the Board: "Given reasonable, intelligent, people, I don't think expertise in special fields is required."

Well, we have just touched on a few of the many inaccurate decisions that the NTSB has made. Now let's examine the talent or lack of it that constitutes the Board.

Mr. McAdams' statement about expertise in special fields not being a prerequisite for a board member is totally in error; however, it does fit the background of the five-member board. Mr. McAdams is the only member of the board who has any flying experience and most of that was gained during his naval tour as a fighter pilot in World War II. From 1958 to 1967 he served as an assistant to a member of the Civil Aeronautics Board. He holds a law degree from Georgetown University Law School.

Oscar M. Laurel was appointed a member of the board in April 1967 by President Johnson. He served two terms as district attorney for the 49th Judicial District of Texas.

Louis M. Thayer was appointed to the safety board in April of 1967 by President Johnson. At the time of his appointment he was retired from the U.S. Coast Guard with the rank of Rear Admiral.

Isabel A. Burgess was appointed to the board in September 1969 by President Nixon. She served on the Arizona Senate for three years prior to coming with the board.

Chairman of the Board John H. Reed was appointed as a board member by President Johnson in April 1967. On May 9, 1969, President Nixon designated him as chairman. Prior to his board service, he was governor of Maine.

The combined aeronautical knowledge of all board members would not come close to the experience of a junior airline co-pilot. Does it seem appropriate to allow a former governor, district attorney, lawyer, admiral, and state senator to damn the dead and to decide the future safety of the traveling public?

It is true that the board receives its technical input from the Bureau of Aviation Safety. The "probable cause" is written by one of its members and checked for potential loopholes by the Office of General Counsel. But the final decision must be made by the board.

John J. Carroll, chief of the Bureau of Aviation Safety, had this to say regarding instrument approaches: "After we made thirteen recommendations on approach and landing accidents we didn't have a repeat of the 1968–69 winter series.

"Look at what these accidents had in common and you'll see that *those things aren't happening any more.*"

Following are a few of the major approach crashes that have occurred since Mr. Carroll's statement:

July 27, 1970. Flying Tigers DC-8.
October 10, 1970. Saturn Airways Lockheed Electra.
November 14, 1970. Southern Airways DC-9.
June 1971. Allegheny Airlines Convair 580.

Yes, we're still having approach accidents and we will continue to have them until something concrete is done.

It would be beneficial to all concerned if the "probable cause" was left to a trial by jury that would listen to bona fide experts. In this way important factual material could be entered in evidence and kept on the records.

In case you think the board's only duties concerning final decisions are restricted to aircraft accidents, you're wrong. Our team of experts are judge and jury for all modes of travel; rail, steamship, highway, and pipeline. Any disciplinary action for licensed pilots, navigators, engineers, mechanics, and dispatchers can be appealed for a hearing by the board. As you see, it's not really a job for a politician.

Board Chairman John Reed said this concerning the NTSB: "We investigate an accident to determine what lessons it teaches and then to make recommendations to prevent a reoccurrence. Our role is not simply to sift through the ashes and announce the probable cause."

Well said. But it has the unreliable sound of a preelection speech.

In June 1972 William R. Haley of Washington, D.C., was appointed by President Nixon to replace Oscar M. Laurel on the National Transportation Safety Board. He is a lawyer with no aeronautical background whatever.

This year the NTSB will become a self-sufficient organization no longer under the thumb of the Department of Transportation, DOT. However, they still have no power to demand safety regulation changes from the FAA. They can only make suggestions which may or may not be heeded.

Chapter Twenty-two

BREACH OF FAITH

ON MAY 21, 1970, President Nixon signed into effect Public Law 91-258, the Airport and Airways Development Act. This act was the legislative response that would enable the aviation industry to catch up and pass its extraordinary growth of the 1960s.

Between fiscal years 1960 and 1969, the total number of aircraft handled by FAA's air route traffic control centers increased by 110.6 percent. The number of planes worked by the towers showed an increase of 112 percent. Federal airport and airway development programs failed to keep pace with the rapidly expanding industry. The answer to these problems is the new Airport and Airways Development Act. This legislation assures a fund of over $11 billion over the next ten years. By establishing a trust fund modeled after the Highway Trust Fund, it eliminates airway and airport development from having to compete for funds, the basic reason for the funding uncertainties of the past.

The money raised by the fund will come directly from the

users of the airways and airports; in essence this appears to be a logical means of obtaining money. Revenues will be generated by the following levies on aviation users: (1) $3 dollar surcharges on passenger tickets for overseas flights originating in the United States; (2) an 8 percent tax on domestic fares; (3) a 7-cent per gallon tax on all fuel used by general aviation; (4) a $25 per year aircraft registration fee on all aircraft, plus a 2-cent per pound (3.5 cents for jets) annual tax for all aircraft that weigh in excess of 2,500 pounds; (5) a 5 percent tax on all air cargo; and (6) the revenues from taxes on aircraft tubes and tires to be transferred from the Highway Trust Fund to the Airport and Airway Trust Fund.

Sections 13 and 14 of PL 91-258 authorize the following expenditures: $250 million per year for air carrier and reliever airports and $30 million for general aviation airports. And not less than $250 million annually for acquiring and improving air navigation facilities.

Section 14 provides: "The balance of the moneys available for the necessary administrative expenses (incident to the airport and airways programs) for the maintenance and operation of air navigation facilities and for research and development activities. It is estimated that the revenues derived from all of these taxes should be over $600 million per year."

If the taxes are used entirely for the purposes for which they are intended, within five years our airports and airways could conceivably have caught up to where they belong. That is a big IF, though. The FAA is trying every trick in the book to divert millions for their personal operating expenses.

The Senate Committee on Appropriations, in its report concerning the supplemental Department of Transportation, said:

> The Committee wishes to make it abundantly clear that the utilization of the Trust Fund receipts the priorities established in the Airport and Airway Development Act be observed. The act clearly makes the funds for administration of the programs the lowest priority and sets forward the funding for airport grants, airway facilities and research and development.

If the FAA budgetary subterfuge is accomplished, $413 million in user appropriations will be siphoned off in the first two years of the bill. This money is intended to support FAA administration and daily operational expenses. Not one foot of runway or an ILS will have been installed.

Senator James B. Pearson of Kansas has charged "breach of faith and doublecross" as a result of the FAA-intended swindle: "Funds earmarked to improve the country's airway system will go instead toward bureaucratic salaries and routine expenses." Pearson stated that the administration plans to spend half the money in Washington for "salaries and erasers at FAA."

And so it goes! A continual fight for survival. One gentleman who has the drive and spunk called for by his position is the new two-fisted president of the Air Line Pilots Association, Captain John J. O'Donnell. In his testimony on June 9, 1971, before the Subcommittee on Transportation and Aeronautics of the House Commerce Committee, O'Donnell pointed to an urgent need to modernize the nation's air transportation system which was the purpose of the ten-year airport and airways legislation passed last year. He posed these questions:

"How would anyone in government explain to the loved ones of the 49 persons aboard Air West DC-9, lost last weekend in California, that a contributing cause to death is a system that will permit two jet aircraft to be on a collision course on different flight plans several miles above the earth?

"How can anybody really explain or excuse the death of 28 persons aboard the Allegheny Airlines prop-jet last Monday that was attempting a nonprecision approach at New Haven, Connecticut, airport?

"The airport manager," O'Donnell emphasized, "stated that 'the crash would not have happened if the airport had an instrument landing system.' "

"During the past ten years," O'Donnell added, "there were 19 fatal and 32 nonfatal airline accidents resulting in 587 deaths and a monetary loss of over $111 million during nonprecision landing approaches. Only 10 percent of the runways provided by the nation's 530 airports used by airline planes have complete ILS systems and other landing aids required for precision approaches."

"To place ILS systems on the remaining unprotected 90 percent of the runways . . . not now programmed for such installations by 1981 . . . will cost $250 million."

He pointed to FAA plans over the next decade to modernize and enlarge its present private fleet of aircraft: "FAA's total planned new aircraft procurement amounts to $123.5 million. . . . This money alone would finance more than half of the ILS systems needed.

"More importantly, such redirected priorities may save the lives of another 587 passengers and prevent loss of another $111 million-plus investment by 1981."

Chapter Twenty-three

WHAT'S TO BE DONE?

THERE'S no cure-all for aviation safety. As long as there is a chance for human error in designing, testing, maintaining, regulating, and flying, the door is always open to possible disaster. We can and must minimize the potential for crashes. I firmly believe that if at least some of the following recommendations were implemented, flying would be as safe as our state of the art will allow. Every passenger that flies on a scheduled airline is entitled to the maximum that the industry has to offer.

Airports. Runways should be at least 6,500 feet in length. There should be ample underrun and overshoot extensions at both ends of the runways and they should be grooved. All runways should have approach lights, centerline lights, high-intensity runway lights, runway-end identification (REIL) lights, and visual-approach slope indicators (VASI). Also essential: a control tower equipped with radar monitoring of all approaches; sufficient fire-fighting equipment to handle the largest aircraft that frequents the field, and backup electric power for all navigation aids, communications, and lights.

Airport security is practically nonexistent; it must be increased. Any person who dons a pair of overalls can get on any parked airliner. The doors are always open to the intruder.

Approach Aids. The only approach that should be allowed at an airline airport is a precision approach. That means that there must be glide-slope guidance for the pilot. An ILS with distance measuring equipment (DME) or a PAR should be the only acceptable letdowns. For maximum safety both should be functioning at the same time so that one could be used as a check on the other. There should be at least two precision approaches on different runways. This would eliminate the necessity of making a circling approach. If the glide slope on the ILS should go off the air during instrument weather and there is no precision radar, that field should be closed. I have flown into airports that have had the glide slope shut down for maintenance for as long as two years. This is a job that the FAA could perform in a few days.

Aircraft. Since the antique "see and be seen" concept of flying is still with us, the pilots of small jets need a third crew member. Strobe lights would greatly enhance the chances of not bumping into another plane. Presently the only required proximity lighting, other than navigation lights, is a small rotating Grimes beacon similar to the ones on police cars. As the barometric altimeter has proved its undependability by killing hundreds of passengers, a radio altimeter is a must. Heated static ports would give the barometric altimeter a fighting chance. Also essential: an automatic pilot and dual distance measuring equipment; standardized cockpits; a warning device that tells you when your instruments are in error, not just when the electric power has been interrupted; a collision avoidance system that will warn a pilot when another plane is on a collision course.

Regulations. There are so many useless rules that should be abandoned. We are regulated to death. It is no longer a case of checking the approach plate and landing. The books must not only be consulted but also studied to see such trivial items as if the sun, moon, and stars are dimly visible or if the cloud cover is thin or normal. There are too many loopholes that enable a flight to land or take off when it really should not. The rules should be simplified and the Air Transport Association's idea of keeping the flight moving at any cost should be stricken from the book.

There is no question in the minds of pilots and controllers as to what is our most pressing problem. It is MID-AIRS and it is getting worse.

Air Traffic Control. Controllers as a rule do a damn fine job. It is a pleasure to watch the artistry of an approach controller at a busy airport as he spaces his traffic and issues advisories until intercepting the final approach. Nevertheless, the FAA should realize that these men are human beings performing an exacting task. No man should be allowed to work a busy sector for longer than four hours. At present they are asked, sometimes ordered, to control traffic when they should be home in bed. You read about how many new controllers the FAA is hiring; what you don't read is that it takes three years to attain journeyman status, and many never qualify. The PATCO newsletter dated June 18, 1971, states: "The FAA hired 2,428 trainees from January 1, 1971 to April 30, 1971. During that time period 1,288 failed to make the program."

The real key to airline safety is found with the men who know flying best: The Air Line Pilots Association, the Professional Air Traffic Controllers Organization, and the Aircraft Mechanics Fraternal Association. Dedicated people such as these take up the long line of slack left by the FAA.

Some of what you have read in this book has probably shocked you. Still, I feel that you, the airline passenger, the one who ultimately pays my salary, deserve the right to know the terrible state that aviation is in. If you are a representative or a senator, please come to the aid of the traveler. If you are not affiliated with the government, a quick note to your congressman will help the cause. You should not have to be called on to make the skies safe. But, in reality, you are the one who can turn the tide.

During 1973, 227 people lost their lives on U.S. carriers. The following is a breakdown of the accidents:

7/73 St. Louis	*Ozark F-227*
7/73 Boston	*Delta DC-9*
9/73 Alaska	*World Airways DC-8*
9/73 Arkansas	*Texas International CV-600*
10/73 Greensborough	*Piedmont B-737*
11/73 Boston	*Pan Am B-707*
11/73 Chattanooga	*Delta DC-9*
11/73 Akron	*Eastern DC-9*

General aviation accidents during this period far overshadowed the airlines. Of 4,215 accidents, 722 were fatal to 1,411 people.

In 1974, the worst year since 1960, there were 467 fatalities on U.S. scheduled air carriers. A chronological list follows:

		Fatalities
1/74 Pago Pago	*Pan Am B-707*	*96*
2/74 Honolulu	*Pan Am B-707*	*1*
4/74 Indonesia	*Pan Am B-707*	*107*
9/74 Athens	*TWA B-707*	*88*
9/74 Charlotte	*EAL DC-9*	*71*
12/74 Berryville	*TWA B-727*	*92*
12/74 Bear Mountain	*NWA B-727*	*3*
12/74 Alaska	*KWA G-21*	*5*

During the same year general aviation had more accidents than in 1973, but the fatalities decreased to 1,290. The world-wide airline death toll for the year 1974 was a staggering 1,815. Certainly not a safety record to be proud of. Many of the accidents could have been avoided with properly equipped airports and a full complement of journeymen controllers.

On January 6, 1974, an Air East Beechcraft crashed while making an approach to Johnstown, Pennsylvania. Twelve people lost their lives in the needless tragedy.

At the NTSB hearing it was revealed that the captain

of the flight was pressured by his company to break the landing limits prescribed by the FAA. It seems that a few chosen captains were cleared by the Vice-President of Operations to descend to lower limits than the law allowed. The captain of the above flight will never again go down to "company minimums."

If there is to be any breakthrough of the safety barrier, certainly immunity from reprimand is a must. Just a few months ago the FAA declared an immunity program for the industry. However they did not specify what they would listen to, and what they would issue violations for.

In essence there is no valid immunity as yet and there must be, if aviation is to survive. If a pilot, controller, mechanic, or dispatcher spots an unsafe situation, he should be free to report it to the FAA without fear of reprimand or dismissal.

The longest accident investigation in history has just been completed. The TWA accident at Dulles in December 1974 caused friction between pilots and controllers. Let us all hope that the loss of 92 lives will not have been in vain.

On February 21, 1975, PATCO President John Leyden put it right on the line in his speech before NTSB:

"Another reason that we are delighted to testify here today is that although PATCO has the responsibility of representing all the nation's air traffic controllers, we have by design been excluded by FAA from participating in any technical area which impacts directly on the system.

"It seems rather strange that participation is solicited from aviation manufacturers, business organizations, and even cabin attendants, and yet there is no voice or input from the individual who is primarily responsible for the day-to-day operation of the Air Traffic System."

When new procedures for air traffic are implemented, the journeyman controller is the man who should be consulted, and not some "managerial type" who hasn't con-

trolled a plane for years. Pilots and controllers should utilize similar charts, approach plates, and regulation manuals. At present the controller's charts are different in many respects from those of the pilot. To cite a case:

The controller's chart for Dulles Airport depicts no mountains west of the field, but the pilot's charts do. The airline pilot has myriad books at his command: charts, regulations, company instructions, let-down plates, aircraft manuals, etc. But what he does not have is a copy of ATP (Air Traffic Procedures 7110 8 and 7110 9, the controller's Bible), and he is not required to even look at these all important manuals.

The working relationship between pilots and controllers is indeed a close one, but the lack of knowledge about the other fellow's job is criminal. Pilots should be made to visit various controller facilities to watch the journeymen at work. Likewise controllers should be able to ride in the cockpit of an airliner whenever there is space for them. Many controllers would welcome this chance, but the FAA only allows a few familiarization trips a year. There is so much red tape involved many controllers just can't fight it.

When an airline decides to add an item to its MEL (Minimum Equipment List), or should I say take one away, who would be the most knowledgable person to consult? The captain, of course. When the MEL boards meet, they consist of FAA and manufactuers' test pilots, representatives of the airline companies, but no line pilots, the ones who have to fly and live with the mistakes of others.

Just as PATCO's forceful president has stated time and time again, the men who know the system best are those who have to keep it going. The same goes for the flying part. When a company wants to change a design or a component on an airliner, it would be highly beneficial to all if the pilot, the man who has to drive the thing, were consulted about just how many components he feels he needs before an MEL is thrust upon him.

The FAA decreed that by December 30, 1973, nearly all non-airline aircraft must be equipped with ELT (Emergency Locator Transmitter). This tiny transmitter activates when exposed to sudden deceleration forces as when a plane comes to an abrupt stop in the side of a mountain. The radio then sends out an automatic emergency signal which can be picked up by another aircraft. The unit is only as large as a milk container and has a two-year self-contained battery in it. There have been numerous false alarms with this unit, but when they finally get all the bugs worked out, I see no reason why the airline passenger could not be afforded the luxury of being found after a crash. These ELTs should be standard on all airlines.

The latest FAA Air Facts Book shows no ASDR (Airport Surface Detection Radar) in operation. I know for a fact that there are units installed at Chicago, Kennedy, and Newark. The controllers at these fields informed me that the radar is operational, but it usually fades out during inclement weather—just when it is needed the most. An ASDR is, as the name implies, a precise radar that is so accurate it will pick up the outline of a plane as it taxis out. This radar is used strictly to control traffic on the surface of the airport when fog-bound. All major airports should have this valuable aid as they once did before the FAA started decommissioning them.

Whenever a Congressional subcommittee is called to investigate the FAA, the aircraft manufacturers, or the airlines, the representative or lawyer who is conducting the investigation should be aided by people in the industry. I have studied numerous transcripts that deal with squandering of money by the FAA, and the Congressmen do a yeoman's job in a field they know little about. But I have seen an investigator build a case on a certain issue only to be tricked by his adversary because of his lack of technical knowledge.

I would like to quote some passages taken from the April 1974 hearings before a Subcommittee of the Committee on Appropriations House of Representatives. Representative Conte is questioning Mr. Butterfield, FAA Administrator, Mr. Moore, Assistant Administrator FAA, and Mr. Belanger, FAA Acting Director Air Traffic, about statements made in Safety Last.

MR. CONTE: *I didn't read the book. My legislative assistant read the book and he says Power-Waters claims controllers are supposed to handle only twelve airplanes and that FAA gives controllers an award for handling thirty-five to forty planes.*

MR. MOORE: *No sir, no way.*

MR. CONTE: *Can we have something in the record on that? This might come up in debate.*

MR. BUTTERFIELD: *I don't mean by my earlier statement to downgrade many of the things the author said. I think FAA and perhaps a number of other organizations can heed some of his warnings with regard to safety generally, but it was pretty obvious to me that he was more than somewhat bitter. That seems very clear. But I don't know about those figures.*

Skipping to the next page of the transcript, we find Mr. Belanger being questioned by Representative Conte who has the proverbial ace-in-the-hole.

MR. CONTE: *We had a hearing on controllers a couple of years ago. I won't mention the name of the individual, but in talking about a number of aircraft he had handled, he said, "I have a letter in my own file, a letter of commendation for having worked thirty-two aircraft at one time. This was sheer chaos."*

MR. BELANGER: *I would say that he did not work by himself; he had one or more additional team members in there. I wouldn't deny that this team could have had thirty-two aircraft, but this is the exception to the rule.*

Up until this point Representative Conte has Belanger's back to the wall, but he wriggled out simply because Mr. Conte lacked the knowledge of a journeyman controller.

Mr. Belanger conceded that thirty-two aircraft could have been worked, but only by a team. That is where he is wrong. No matter how many men are helping a controller, he and he alone has the microphone in his hand. He alone is the person who is talking to the pilots, therefore he is the only one controlling thirty-two planes.

Not wishing to downgrade in any way the tremendous job done by the representatives, I would simply like to give them a technical back-up. If a journeyman controller had been sitting by Mr. Conte's side and told him that only one man works the radio, what would Mr. Belanger's answer have been then?

I finally believe that a representative should have a skilled technician with him to aid him in ferreting out the truth. If the ATC system is under fire, a PATCO controller is needed. If it's engineering, a skilled AMFA mechanic is the answer. If it's a flying problem, an ALPA air safety pilot is required.

All radio navigational identifiers should be converted to tapes that utilize voices rather than Morse code. It just goes to show that the FAA thinking is still in the Morse code era. A voice identifier takes the workload off the pilot. An airline crew is busiest in the landing phase, and they should not have the added burden of distinguishing dits from dahs.

Large metropolitan airports should provide separate runways for light aircraft. Trying to keep proper spacing behind a light plane traveling 50 knots slower than you is a thankless task.

With the energy crunch many thousands of gallons could be saved by averting the go-arounds that result when the little fellow does not clear the runway in time. An average go-around takes 6 minutes of additional flying time. The twin jet burns approximately 75 pounds of fuel each minute. Therefore each go-around consumes 450 pounds of fuel. A conservative estimate of the amount of fuel wasted on go-arounds is 632,396 gallons per year. It is also far safer for the light plane to land on a strip where he won't have to fight the treacherous wake left by the jets.

Installing more ILS will also save fuel as there would be fewer missed approaches requiring a flight to continue to his alternate airport.

The no man's land—that space from the top of the TCA to the bottom of the positive control airspace at 18,000 feet—still exists for the air traveler. There are some 10,000 feet of airspace in which aircraft fly unbeknownst to the air traffic controller. The height of the TCA in Atlanta has recently been increased from 8,000 feet to 12,500, a step in the right direction, but it still leaves 5,500 feet of unprotected airspace. The airline passenger is entitled to the safest flight possible. The mid-air potential will not be minimized until the top of the TCA is raised to meet the positive control area or the positive control area is lowered to the top of the TCA. This is not an unreasonable request even though general aviation pilots will scream that they are being regulated out of the sky. Canada has long utilized the above recommendations. All of the TCA in Canada (there they are called TESA terminal radar service area) extend to 9,500 feet where they meet the safety of positive control airspace. An airline passenger in Canada is under positive radar control from takeoff to landing and has been for years. Why then can't the U.S. follow suit on such a life-saving procedure?

If you are interested in promoting a safer system and have written to your Congressman with little result, please contact Ralph Nader's Aviation Consumer Action Project, P.O. Box 19029, Washington, D.C. 20036. Mr. Nader has done a great deal to promote aviation safety and he needs your support to continue his crusade for safety in the sky.

GLOSSARY

AILERON: the movable part on each wing of an aircraft that allows the pilot to bank the plane.

AIR TRAFFIC AREA: a five-mile radius, up to 3,000 feet, from the center of an airport.

AIRPORT SURFACE DETECTION EQUIPMENT (ASDE): a radar system that detects traffic movement on the ground in bad weather or when visibility is poor.

AIRPORT SURVEILLANCE RADAR (ASR): a type of radar that has no height-finding capabilities but is directional only.

AIRSPEED AND MACH INDICATOR: an instrument that tells the pilot his speed through the air—vital if he is to avoid exceeding maximum speed or stalling while attempting to land.

ALPA: Air Line Pilots Association. The commercial pilots' union.

ALPHA NUMERICS: a radar system to give the air traffic controller the altitude of his traffic as well as its direction.

ALTIMETER: an instrument for measuring altitude. Three-pointer altimeters, though dangerously susceptible to misreading, are still in use. Single-pointer drum-type altimeters are an improvement in that they reduce margin for error, but digital-readout altimeters (measuring altitude from 0 to 1,000 feet, and from 1,000 to 50,000 feet at a single readout) are the safest now in use. Altimeters based on radar are far more accurate than those based on barometric pressure.

ALTITUDE-ALERTING SYSTEMS: buzzers or warning lights that alert the pilot when he is approaching assigned altitude.

APPROACH: getting the plane safely through the weather and onto the runway. Various approaches (from worst to safest) include radio beacon, circling, visual oral range (VOR), airport surveillance radar (ASR), instrument landing system (ILS), and precision approach radar (PAR). See listings under each approach.

APPROACH PLATES: diagrams showing approach altitude, proper course headings, time in minutes (or seconds) from approach facility until you can expect to see runway. Descent limits are based on type of approach to be used and the terrain.

APU: see AUXILIARY POWER UNIT

ARTIFICIAL HORIZON: invented in 1928, this is a gyro that enables the pilot to keep a miniature horizon in front of him at all times. See FLIGHT DIRECTOR for an improvement on this system.

ASDE: see AIRPORT SURFACE DETECTION EQUIPMENT.

ASR: see AIRPORT SURVEILLANCE RADAR.

ATA: Air Transport Association. An organization of commercial scheduled airlines.

ATTITUDE: the raising or lowering of the nose of the aircraft in relation to the horizon.

AUTOPILOT: automatic pilot; a device for steering an aircraft automatically.

AUXILIARY POWER UNIT (APU): contained in the tail of most aircraft, this unit supplies electricity, pneumatics, and air conditioning when the aircraft engines shut down.

BEACON APPROACH: see RADIO BEACON.

CAA: Civil Aeronautics Administration. Has been replaced by the FAA.

CAB: Civil Aeronautics Board. A government agency primarily concerned with routes and tariffs.

CEILING: altitude of lowest cloud deck. Broken or overcast.

CHECK LIST: the list of items to be read aloud by first officer and responded to by captain on takeoff and landing. A full reading and response is mandatory on all commercial flights.

CHECK RIDE: a flight on which an FAA inspector or an FAA-authorized pilot accompanies the pilot and grades him on his overall flying ability. All line pilots are required by the FAA to pass check rides yearly.

CIRCLING APPROACH: when the center line of the runway and the navigational approach aid differ by more than 30 degrees, a circle must be made to land. The pilot descends to approach minimums, sights the field, and turns the ship downwind. He gauges his landing solely on directional runway lights and ground visual cues; a primitive approach method.

CLEARANCE: permission to take off, proceed to a specified altitude, change course, or land. Clearance is given from control tower (or radar center) controlling area's flights.

CONTROL ZONE: a block of protected air space around one or more airports.

DISTANCE MEASURING EQUIPMENT (DME): A navigational aid that tells the pilot his exact distance from a VOR station on the ground. (See VOR.)

EFFECTIVE RUNWAY LENGTH: the total usable portion of the runway, allowing for safe margin of altitude over approach obstructions.

ELEVATOR: the component of the aircraft located in the tail section that controls the up and down movement of the ship.

FAA: Federal Aviation Administration. The government agency responsible for flight safety.

FAR: Federal Air Regulations. Laws governing flight, made up and enforced by the FAA.

FIRST OFFICER (F/O): the co-pilot.

FLIGHT DIRECTOR: an attitude indicator that depicts the horizon, the relative position of the aircraft, and the pitch of the ship. It displays every movement the aircraft makes—up, down, left, or right.

F/O: see FIRST OFFICER.

G (GRAVITY): a unit of force equal to the force exerted by gravity on a body at rest and used to indicate the force to which a body is subjected when accelerated.

GCA (GROUND CONTROL APPROACH): see PRECISION APPROACH RADAR.

GENERAL RAILROAD SIGNAL DEVICE (GRS): a device used in railroad freight yards which can also be used to signal between airport control facilities when there is no interphone between facilities. The GRS, so used, is primitive and dangerous.

GLIDE PATH: a signal from a radio transmitter approximately 1,000 feet down and to the side of the runway which guides the plane to the landing strip at an angle of 3 degrees, depending on local terrain. Part of ILS system.

GRIMES BEACON: a small, rotating navigational light, similar to the rotating lights on police cars.

GROUND CONTROL APPROACH (GCA): see PRECISION APPROACH RADAR (PAR).

GRS: See GENERAL RAILROAD SIGNAL DEVICE.

HYDROPLANING: the sliding that occurs when there is a thin film of water on the runways separating the aircraft's tires from the surface, making braking impossible.

IDENT: a button in an aircraft connected to the transponder that when pushed, sends a pulse to the radio controller which positively identifies the aircraft by nearly doubling the size of its radar blip. (See TRANSPONDER.)

INSTRUMENT FLIGHT RULES (IFR): rules governing aircraft operation when weather is bad and ground contact cannot be made. *All* commercial turbojets must use IFR regardless of the weather. IFR gives protection of controlled separation of commercial aircraft, normally a minimum of 1,000 feet vertically when flying assigned routes.

INSTRUMENT LANDING SYSTEM (ILS): an approach system enabling the pilot to align aircraft with runway entirely by instruments. Component parts are localizer, glide path, outer and middle markers (see separate listings). Its effectiveness depends on proper functioning of all components; it is also affected by weather.

JOINT ACCEPTANCE INSPECTION TEST (JAIT): test that all equipment must pass before it can be accepted by an airline.

LANDING FEE: the amount that each aircraft must pay in order to land at a commercial airfield; it is based on the weight of the aircraft.

LAYOVER: a flight on which the pilot does not immediately return to his base station (see TURNAROUND).

LINK TOWERS: microwave towers spread 30 miles apart that pass radar signals from one tower to the next until they can reach the radarscope of the air traffic controller. Above ground, they can be affected by weather or vandalized.

LOCALIZER: a radio transmitter placed on the far end of the runway to give the pilot directional guidance between 3 and 6 degrees. Part of ILS system.

LOGBOOK: the record of a specific aircraft's exact time for each flight; length of time airborne of the airframe (skeleton of the plane), engines, and components; mechanical discrepancies noted by pilots and mechanics.

MEL: see MINIMUM EQUIPMENT LIST.

METAL FATIGUE: a structural failure in the aircraft body, often caused by high local stress concentration. Cracks in the wing section or fuselage are telltale clues of metal fatigue.

MIDDLE MARKER: a radio transmitter that alerts the pilot as to his distance from the field. It activates a flashing light and an aural signal in the cockpit usually 3,500 feet from runway threshold. Part of ILS system.

MINIMUM EQUIPMENT LIST (MEL): written by the aircraft manufacturers and approved by the FAA and ATA, the MEL lists various components on the aircraft that must be operative for minimum safety.

NTSB: National Transportation Safety Board. A federal agency which investigates the causes of aviation accidents and makes recommendations to the FAA for safe practices.

OUTER MARKER: a radio transmitter that alerts the pilot as to his distance from the field. As aircraft passes over outer marker a colored light flashes in cockpit and an oral signal is received. Outer marker is usually placed 4 to 7 miles from end of runway. Part of ILS system.

PAR: see PRECISION APPROACH RADAR.

PATCO: Professional Air Traffic Controllers Organization.

PRECISION APPROACH RADAR (PAR): an accurate radar guidance system for landing, PAR affords both direction and altitude guidance, with the guiding voice from the final controller. System is also called "Ground control approach (GCA)."

RADAR BOARD DISPLAY EQUIPMENT (RBDE): a set of components including screens and boards that monitor radarscopes for air traffic control. To be effective, units must match radar units exactly.

RADAR HANDOFF: a system by which an air traffic controller relinquishes his guidance of a flight after a receiving controller has the flight's "blip" positively identified on his radarscope.

RADIO BEACON: originally developed in 1930, this low-powered low-frequency radio enables the pilot to home in on the airport radio station. It is nondirectional and affected by weather and geographical conditions; an unreliable radio-landing aid.

RUNWAY CONDITION READING (RCR): a device that provides accurate braking-action reports on a specific runway so that the pilot can plan a safe landing approach.

RUNWAY VISUAL RANGE (RVR): the number of feet a pilot can expect to see down the runway when he breaks out of the clouds.

SIMULATOR: a ground-school training aid; a mock-up of the cockpit of an aircraft that can simulate in-flight conditions while remaining on the ground.

SPOILERS: sections of top of the aircraft's wings that extend to reduce speed on landing.

STRAIGHT-IN APPROACH: under instrument flight rules (IFR), an instrument approach wherein final approach is begun without first having executed procedure turn (see CIRCLING APPROACH).

STROBE LIGHTS: electronic lights mounted on each of the aircraft's wing tips (similar to a camera flash gun). Especially useful in hazy crowded skies near large airfields.

SUPPLEMENTAL AIR CARRIERS: nonscheduled airlines.

TECHNICAL STANDARD ORDER (TSO): material that has the approval and certification of the FAA.

TRANSPONDER: an airborne device that automatically receives radio pulses directed to it by the air traffic controller. When device is activated by a controller it triggers the airborne transmitter, which gives a two-slash blip on the controller's radarscope. If pilot presses his Ident button (q.v.) the size of the blip is doubled.

TSO: see TECHNICAL STANDARD ORDER.

TURBOJET AIRCRAFT: planes that are not propeller driven, i.e., prop jets.

TURNAROUND: a flight on which the pilot takes the aircraft to a specified airfield and returns to his base station. See LAYOVER.

UNDERRUN AND OVERRUN: the additional paving that is on each end of the landing strip.

v1: decision or commit speed; the speed after which takeoff must be continued.

v2: the minimum climbing speed after takeoff; recorded in the manifest for each flight.

VASI: see VISUAL APPROACH SLOPE INDICATOR.

VERTICAL SPEED GAUGES: gauges that measure velocity up or down in hundreds of feet per minute.

VERY HIGH FREQUENCY OMNI DIRECTIONAL RANGE (VOR): the most widely used method of radio navigation. The VHF radio signal emitted is supposedly unaffected by static and a ground transmitter projects a possible 360 courses. Drawbacks: reception range is governed by line of sight; weather affects its performance; there are a limited number of VOR frequencies; it gives direction only, not descent angle. Best used with DME (see DISTANCE MEASURING EQUIPMENT).

VISUAL APPROACH SLOPE INDICATOR (VASI): a series of colored lights that tell a pilot when he is in the proper approach path for landing.

VISUAL FLIGHT RULES (VFR): non-instrument flying used in good weather; criteria are 1,000-foot ceiling (lowest cloud deck) and 3-mile visibility. Regardless of weather, however, all commercial turbojet flights must use IFR (see INSTRUMENT FLIGHT RULES).

VISUAL ORAL RANGE (VOR): the most widely used method of radio navigation. The VHF radio signal emitted is supposedly unaffected by static and a ground transmitter projects a possible 360 courses. Drawbacks: reception range is governed by line of sight; weather affects its performance; there are a limited number of VOR frequencies; it gives direction only, not descent angle. Best used with DME (see DISTANCE MEASURING EQUIPMENT).

VOICE RECORDER: a device that tape-records conversations in the cockpit of an aircraft. The voice recorder has been used effectively in the reconstruction of accidents.

WAKE TURBULENCE: the wind disturbance caused by aircraft as they take off and land. The "wake" behind a 707 has the velocity of a hurricane; that of a 747 is 10 percent higher. On calm days, turbulence may linger behind an aircraft for as long as 5 minutes.

WEATHER RADAR: a type of radar on an airplane that can indicate the best route around storms; it does *not* pick up other aircraft on its scope, however.

INDEX

Author's Page

Captain X (Brian Power-Waters) has been an airline pilot for more than 20 years, and has logged more than 23,000 hours in the air. He still spends 85 hours a month piloting jets for a major airline.

A veteran and qualified fighter pilot with command pilot wings, he is a major in the USAF Reserve. For three years he was also in the Royal Canadian Air Force, and is a member of the RCAF Reserve Association. He still holds his 20-year's rating as a military pilot and also has air transport ratings for a variety of aircraft.

He is rated as an instrument instructor qualified to teach everything from gliders to jets. He is also a flight instructor for seaplanes and a ground instructor in meteorology, radio navigation, civil air regulations, and aircraft engines. He is a member of the Parachute Club of America, the Experimental Aircraft Association, and the International Aerobatics Club. He has a hot-air balloon rating, is a senior parachute rigger (licensed to pack parachutes) and belongs to the Aviation Space Writers Association and the Aerobatic Club of America.

Captain Power-Waters was born in London, and now makes his home in Church Hill, Maryland.

THE AVIATION SAFETY INSTITUTE

Safety Reporting Form

You may use this special form to report any unsafe act or condition you observe while flying or otherwise involved with a passenger-carrying aircraft. You are not required to identify yourself. Please be as specific as possible regarding conditions and personnel or equipment. If you need more space to complete your report you may use the back of this page, attach a letter, or call ASI directly at (614) 885-4242.

We are here to help you make flying safer.

MAIL TO: THE AVIATION SAFETY INSTITUTE
Safety Reporting Section
P.O. Box 304
Worthington, Ohio 43085